C000318664

A Garden Bird Year

A Garden Bird Year

A MONTH-BY-MONTH GUIDE TO UNDERSTANDING YOUR BIRDS

MIKE TOMS

WILLIAM COLLINS

William Collins
An imprint of HarperCollins*Publishers*
1 London Bridge Street
London SE1 9GF

WilliamCollinsBooks.com

HarperCollins*Publishers*
1st Floor, Watermarque Building, Ringsend Road
Dublin 4, Ireland

First published in Great Britain by William Collins in 2021

2022 2024 2023 2021
2 4 6 8 10 9 7 5 3 1

A catalogue record for this book is
available from the British Library

ISBN 978-0-00-847061-6

Typeset in Berling by Jacqui Caulton
Printed and Bound in the UK using
100% Renewable Electricity at CPI Group (UK) Ltd

MIX
Paper from
responsible sources
FSC™ C007454

This book is produced from independently certified FSC™ paper
to ensure responsible forest management.

For more information visit: www.harpercollins.co.uk/green

To Annabel Hill,
who showed me the meaning of home.

Contents

Introduction

Time spent at home shapes our perceptions and changes the ways in which we interact with the wider world. As we discovered during those initial months of the coronavirus pandemic, increasing time spent at home leads you to notice the small things around you, to take greater interest in your garden and its wildlife, to make home and the creatures with which you share it the focus of your world. Busy lives mean that we have lost a lot of that focus, but maybe by maintaining an interest in garden birds we can get some of that back.

Garden birds provide a window onto the natural world: the behaviours that they exhibit mirror those seen in other creatures, living in other habitats or in different countries. By understanding their lives, we can better understand the lives of creatures that we only ever glimpse through the medium of television or online. If we learn to appreciate our garden birds and to recognise them as fellow living creatures, then perhaps we can learn to take more interest in the lives of those others, living at a distance in the jungles of Southeast Asia or on the plains of South America. Our activities have an impact on a global scale, through the clothes we buy, the foods we eat and the consumables with which we fill our homes. If we can foster new relationships with our closest neighbours – the birds in our gardens – then perhaps we can learn to take greater care of these other natural communities, wherever they occur.

I think it is unlikely that you are the kind of person to simply put out food for visiting birds and then take no more interest in them. In fact, I think that very few people fall into that particular category. We are a naturally curious species and the sights and sounds of garden birds are inevitably of interest. We delight in seeing them visit, relish in watching their behaviour, and gain a warm feeling from the thought that, by providing food, we are helping them to get through the winter. The provision of food feels to us like a helping hand, that we are in some way atoning for our wider impacts on fellow creatures. It is an act of charity, a generous provision, and something that binds us to the natural world around us. Of course, it isn't all about the birds; in most cases it is as much for our benefit as theirs. Just as the recovery times of hospital patients have been shown to improve from having a view of the natural world, so the view onto our bird tables and hanging feeders provides interest, a sense of connection and nourishment for our health and well-being.

This book is about the birds that visit our gardens. It is about why they visit and what they do while they are here. It provides an opportunity to see these visitors in new ways, to recognise who they are and to understand their behaviour. The view from your kitchen sink or comfortable chair provides a window, quite literally, onto the world of these birds and brings you closer to their lives. As you discover more about them you will, inevitably, look more closely and take more interest. These will become 'your' garden birds and you will begin to recognise them as individuals, each one making its way through the world. You will see different behaviours as the months slip by and the seasons change, from the arrival and departure of winter visitors, to the appearance of a new generation of young birds during the summer.

I have been fortunate enough to have watched the birds in a number of quite different gardens over the years, and to have studied them professionally, all the while adding to my knowledge and enjoyment of them. I have been able to delve into the research papers that document our growing understanding of their behaviours, and to have participated in studies that have answered questions about the ways in which birds use gardens and the resources that they provide. It is through this work that I have come to view garden birds in a new way, to appreciate their adaptability and the extent to which they have worked the food, nesting sites and other opportunities that we provide into a wider set of resources, many of which lie far beyond the borders of our gardens. Such findings challenge the old notion that the birds using our gardens are commonplace, unexciting and far from special. They may be familiar in the sense that we recognise them, but this familiarity is superficial and there is much still to learn. What we have learned so far will change how you view your garden birds and I hope that it will also inspire you to help increase our knowledge further, so that we can better look after these amazing creatures.

January

January feels like a good place to start this book, not just because it is the start of the calendar year but also because it is a month filled with optimism and the chance to begin something new. It is the month of resolutions, when many of us first put out food for visiting wild birds or take the opportunity to participate in 'citizen science' through the widely promoted Royal Society for the Protection of Birds (RSPB) Big Garden Birdwatch. It can also be a particularly rewarding month for the garden birdwatcher, with plenty of birds visiting the garden and hanging feeders very much the centre of activity. It is true that the daylight hours are short, that the weather can prove challenging, but these too provide the focus needed to take more interest in your garden birds. It is during the winter months that home becomes more prominent in our lives, the outside world seen increasingly through a window from the warmth and comfort of a kitchen or living room. We range less widely and have fewer hours of daylight available to us, shortening our horizons and the opportunities to engage with nature and the countryside.

At the heart of our relationship with garden birds is food. Food is a key resource and one whose presence appears to be central in determining whether or not birds will make use of your garden. Put up a bird feeder with the right food, perhaps mixed seed or sunflower hearts, and it will not be long until it is discovered by the local tits, finches and sparrows. If yours is a new garden, perhaps a largely blank canvas on a new housing development, then it may take a while for birds to find your feeders. This is likely to be because the garden lacks the other features, such as trees and shrubs, that act as cues for small birds and attract them to the general area. If your garden sits among other mature gardens, well endowed with bird tables and hanging feeders, then you are likely to start receiving interest within a few hours of your new feeders going up. Either way, as we will discover, if you provide the right food in the right way then the birds will come.

More of us provide food for wild birds in the winter months than at other times of the year, and the practice of putting out a few kitchen scraps almost certainly has its origins during this season, when poor weather conditions bring wild creatures closer to our homes. A growing number of us now provide food year-round and would worry about our garden birds were we to miss a week. Year-round feeding is championed by the big conservation organisations like the RSPB and the Wildlife Trusts; it is also supported by the research carried out by the British Trust for Ornithology (BTO). The provision of food has now become part of a wider practice of wildlife-friendly gardening and for many homeowners the two now go hand in hand. Of course, the community of birds visiting the garden shifts with the season, as does their behaviour, and this provides an ever-changing portfolio of things to see.

The pattern of arrivals at the garden feeding station is not random; it changes from day to day depending on the weather conditions and it also differs between the individual species that visit. Get up early enough, ideally before dawn, and watch your bird feeders. You may notice that some birds tend to arrive as soon as it gets light, while others do not arrive until much later in the morning. While it is impossible to say whether the observations from a single garden truly reflect patterns of bird behaviour occurring more widely, it is possible to identify common behaviours through the use of 'citizen science' projects, run at the national level and involving contributions from many thousands of gardens. The results from such studies put your own observations into context, marking them out as commonplace or perhaps unusual.

One citizen science project, the Shortest Day Survey – run in December 2004 by researchers at the BTO and working with the BBC Radio 4 *Today* programme – set out to see if there was a pattern to the time at which particular species arrived at garden feeding stations on a typical winter morning. Nearly 6,000 people took part in the survey, their observations revealing that certain species tended to arrive first and others to arrive later. The first to arrive, blackbird (on average arriving 15.4 minutes after first light) and robin (on average arriving 19.9 minutes after first light), were characterised by having large eyes relative to their body size, suggesting that they were able to start foraging while it was still too dark for other species to venture out. The last of the common species to arrive were greenfinch and starling; on average, these two species did not turn up at bird tables and hanging feeders until 45 minutes after first light. When you think about it, these results make sense. A larger eye lets in more light, and having a larger eye relative to your body size should increase your ability to see at lower levels of light. If you can't see that well, then you are not going to risk venturing out.

The results from the study also revealed another interesting pattern, namely that rural populations of a species tended to arrive at garden feeding stations earlier than urban ones. At first glance this seems counterintuitive; you would expect urban environments to be better illuminated than rural ones because of the street lighting that is a feature of our towns and cities. Surely this should make it easier for birds to see what they are doing, enabling them to leave their overnight roost sites that much earlier in order to find food. However, there is another important difference between urban and rural habitats, and that is temperature. The heat escaping from our homes, shops and offices increases local temperatures within urban environments, elevating them above what they would otherwise be. In our largest cities, local temperatures may be six or more degrees above those in the surrounding countryside; this is known by scientists as the 'urban heat island effect'.

Since small birds can use a significant portion of their energy reserves simply keeping themselves warm overnight, any additional temperature boost should be welcome, reducing the amount of fat reserves that they have to burn. If it is warmer overnight, and small birds burn off less of their reserves, then presumably there is less urgency to find food in the morning. On any given winter night, a blackbird roosting in a London park is likely to face less challenging temperatures than one roosting in some blackthorn scrub on the South Downs. This being the case, the London bird is likely to dip less deeply into its fat reserves and be under less pressure to replenish them come first light. Could this be the reason for the difference seen between the average arrival times of urban and rural populations of the same species?

Quite possibly, but more work is needed if we are to untangle the seemingly opposing effects of light and heat, and to properly account for other factors, such as roosting behaviour and competition between species. Some birds may have to travel farther than others in order to exploit the food being provided at our bird tables. While blue tits and wrens may roost in garden nest boxes, other species might not be able to find the roosting opportunities they require within a garden or elsewhere in the wider urban environment. In such cases, it may be that some of the birds using your garden feeders each morning will have roosted farther afield, perhaps in farmland hedgerows or scrubby woodland cover.

Because birds are mobile, their roosting and feeding sites do not necessarily have to be close together. When you think about it like that, you get a sense of the degree of connection between our gardens and the wider countryside. Once you throw into that mix the knowledge that some of our wintering finches will have arrived from Norway, or blackbirds from Germany, you begin to see that these connections reach out much farther. Your garden does not exist in isolation; for the birds at least, it is one resource opportunity in a much grander landscape. Just as the birds visiting your garden on a January morning might be passing through, before moving on to feed at your neighbour's table, so across the year they may take in a broader suite of habitats and feeding opportunities.

One of the questions I am often asked about feeding garden birds is the degree to which individuals might have become dependent on the food that we provide. Some people worry that if they were to stop feeding, it would have a significant impact on 'their' birds – those individuals that they believe rely on the seed and peanuts put out each morning. Because birds are mobile they can move elsewhere, finding other gardens where food is being provided or using the more natural feeding opportunities present in other habitats. Of course, it remains true that the very significant amount of food that we collectively provide for small birds has almost certainly had an impact on their populations and behaviour, but at the scale of the individual garden this is insignificant. The late Chris Mead, a former colleague and the person who taught me to ring wild birds (attaching small metal rings to their legs so their fate and future movements can be tracked), used to say that if you saw four or five blue tits on your hanging feeders on a winter morning, then you would almost certainly have many dozen different individuals passing through your garden during the course of the day. Chris was able to draw on his experience of catching and ringing blue tits in his rural garden, the figures confirmed by the numbers of different individuals he would catch visiting his feeders during the day. Of course, Chris's garden was rural and he provided a lot of food, but such numbers provide an indication of how tits and other small birds move around an area in search of feeding opportunities. If you think that you have an exclusive relationship with your garden birds, then think again.

These patterns of movement change throughout the year, the most noticeable difference being between the breeding season (when birds are settled on a territory) and the winter (when they can range more widely in search of food). In some farmland species, such as goldfinch, there is evidence of a movement out of farmland and into gardens as the winter progresses, while larger-scale movements – such as those between the Continent and UK gardens – may be driven by changing weather conditions or the availability of favoured seed and berry crops across a much larger area. We'll come back to this later. While there is some consistency in the pattern of movements seen within a species, it is worth noting that there are often differences between populations. Many of our breeding thrushes and finches are resident in the UK throughout the year, but their numbers are swelled in the winter months by the arrival of others of their kind from populations breeding farther north and east. That our birds are resident and others are migratory reflects the different weather conditions and winter feeding opportunities experienced in the areas inhabited by these populations. The farther north and east within Europe that a bird breeds, the more severe the winter; faced with such conditions, many small birds are forced south and west during the autumn in search of a less demanding environment.

As we've just noted, some of these movements are weather dependent, but others are driven more directly by food availability. The size of seed and berry crops can drive the irruptive movements of waxwing and the more regular movements of chaffinch and brambling. Even within a population there may be differences. In the case of Scandinavian chaffinches, for example, the females tend to move a greater distance than the males, and so typically winter farther to the south and west. This pattern of movement – termed 'differential migration' – led Carl von Linné, the Swedish botanist and founder of modern systematic botany and zoology, to give the chaffinch the scientific name *coelebs*, meaning 'bachelor' and reflecting the strong male bias to the populations he encountered in Sweden during the winter months.

Our goldfinch population provides another example, in this case with an interesting mix of migration strategies evident. As the activity at your January bird table will probably tell you, many of our breeding goldfinches remain here for the winter, either staying close to their breeding sites or perhaps moving a little farther south, though still remaining within our shores. Other individuals move to the near Continent, wintering in Belgium or the Netherlands, while some of our goldfinches winter in western France or Spain. Sex ratios in the goldfinch populations wintering in these different locations indicate that females move farther south than the males. There is also evidence from bird ringing that a goldfinch wintering overseas in one year might choose to remain in the UK the following one. Such variation in migration strategy often reflects variation in food availability, either directly (is there enough food available here?) or through an individual's competitive status (can I gain access to the food that is here?). Local weather conditions can also be a strong influence on food availability; snow on the Continent will often force ground-feeding finches and thrushes to cross the English Channel in search of food, and under such conditions we typically see an influx of birds. While there is still much to learn, it is important to remember that there is a lot more to the birds using your bird table and hanging feeders than necessarily meets the eye. That wintering chaffinch could have come from Sweden, that blackbird from Poland, and next January that goldfinch might be wintering in Spain.

* * *

It is the provision of seed and suet-based products, coupled with additional foods like mealworms, and water, that attracts many birds into our gardens. This food represents a significant resource and it seems that it is a resource that we are very happy to supply. While some of us pay little attention to the type of food we provide, thinking that 'bird food' is 'bird food', most of us take time to select those foods that are the most appropriate for the birds that visit. We recognise that a food favoured by one species may not be favoured by another, or that food presented in a particular type of feeder might be used by one species but not by others. Such choices in food selection reflect how and where particular birds prefer to feed and the types of foods that they favour. It is knowledge of these preferences that can help you to maximise the potential of your garden for visiting birds and increase your enjoyment of those that come.

Look at any bird closely and you'll get an idea of what it is likely to eat and where it is likely to feed. Take a robin, for instance: although relatively small in size, it has large eyes, long legs and a fine bill. It feeds on small invertebrates and small seeds (hence the fine bill), taken mostly from the ground (the long legs) by active searching (the large eyes). Now take a different species, the blue tit. This has a slightly more robust bill (shorter and stouter), shorter legs but strong grasping feet. Again this is an active forager, but it feeds in trees and shrubs, plucking insects and spiders from leaves and branches or switching to larger seeds in the winter months. Many people confuse house sparrow and dunnock, the latter sometimes wrongly called a 'hedge sparrow', but look at the bill of each species. In the house sparrow it is large and robust, useful when dealing with grain and big seeds, while that of the dunnock is fine in structure. Like the robin, whose bill is very similar, the dunnock feeds mostly on insects, spiders and small seeds. Unlike the robin, however, the dunnock prefers to feed under cover, moving through the vegetation in a mouse-like manner, while the robin feeds more in the open. Just by looking more carefully, you can begin to unravel the behaviour of these birds and to understand the differences between them.

With so much going on at the bird table and hanging feeders during January, it can be difficult to know where to look. The succession of visiting birds can mean that each remains on the feeders for just a few moments, soon to be ousted by another individual. Competition is often less fierce on the bird table or on the ground below, where fallen seed is taken by ground-feeding finches, dunnocks and robins. Regardless of where the birds are feeding, start by getting an idea of which species are visiting. Over time you will build up a list of the regular visitors and you can use this list to direct your future watching, perhaps as you look out for a certain behaviour or seek to identify a particular plumage. You can also use this list to get a sense of which species you might be missing. If you live near a wood, but never see nuthatches or great spotted woodpeckers visiting your garden, could this be because you are not providing peanuts in a mesh feeder? If you don't have young great tits using your hanging feeders in the summer months, could this be because there are no nesting opportunities nearby? If so, then the provision of a suitable nest box could be the solution.

It is not just what you do in your garden that will determine which species you attract. This is also determined by other factors, typically over which you have little or no control. No matter what you do or which foods you provide, you are never going to attract a cirl bunting or crested tit to your suburban garden located in leafy Surrey. Both of these species have a restricted distribution – the cirl bunting to parts of southern Devon and the crested tit to the Highlands of Scotland – so they will only appear on your garden list if you happen to live in one of these two areas. More broadly, species tend to be found in or close to the habitat that they favour for breeding. If your garden is rural in nature, then you have a much better chance of attracting farmland species like tree sparrow, yellowhammer and reed bunting than someone living in the suburbs. Live close to broadleaf woodland and you are more likely to get great spotted woodpecker and treecreeper; live close to coniferous woodland, then siskin is more likely. Since 80 per cent of us live within towns and cities, and 40 per cent of us live in London or the handful of other major conurbations, most of us will see a similar suite of species visiting our gardens. This suite will be reduced during the breeding season, when birds are tied to their breeding habitats, compared to winter, when birds can range more widely. Some birds, such as summer migrants or winter visitors, will only be seen at certain times of the year.

As we have just seen, where your garden is located is important, as is the nature of the habitats around it, and both contribute to the pool of species that will be present locally and potentially able to visit your garden. While you have little or no control of these – save for moving to a new house or helping to shape the management of nearby habitats – you do tend to have control over your garden. As well as providing food in the form of hanging feeders and bird tables, you can also plant and manage the garden to be as attractive as possible to visiting birds.

Taller trees and shrubs, either within your garden or close by, are often used by birds as they arrive. A small flock of finches or thrushes will often drop into the upper branches of nearby trees or shrubs before working their way down to the feeding station. Look at how your garden sits within the landscape and ask yourself where planting some shrubs, or perhaps a larger tree, might improve how your garden connects to this wider landscape.

Think about the garden itself, about where the bird table and hanging feeders – your garden 'feeding stations' – are positioned within it. Is there tall cover close to the feeding station, so that birds can feed in confidence and dive into cover if a predator appears? Or are the feeders left exposed, hanging from the washing line in the middle of the lawn? Is there an obvious route to the feeders? Are they too close to the house, or positioned too close to a large window or other obstacle that might prove a hazard for a visiting bird? Are the feeders positioned just for your benefit, rather than taking into account the needs of the birds? All of these are things over which you have control, so think about what a bird might want and then plan the placement of the bird tables and hanging feeders accordingly. You also need to take into account other wildlife-friendly measures that you might adopt. For example, it is not a good idea to place a feeding station close to a nest box; if you do, the birds breeding in the box will spend valuable time chasing other birds away from their territory, time that should instead be spent on finding insects and spiders for their growing chicks. And certainly never buy a bird table that has a nest box built into its roof! There are plenty of places where you can get good advice about feeding and looking after visiting birds. While the text of this book discusses much of this, for detail on what you can do – from the best bird-friendly plants to the dimensions for nest boxes – there are other, more detailed resources.

* * *

Mid-month, as the New Year resolutions begin to bite, I find that the drop in temperature is being matched by an increase in the amount of food taken from the hanging feeders. Goldfinch and greenfinch dominate on those feeders filled with sunflower hearts, leaving the mesh feeder and its peanuts to the visiting blue and great tits. The strong feet of the tits enable them to cling to the mesh surface, something that the finches would find more difficult. Competition for the food is fierce, especially during the first hours of the morning, as birds jostle to access the circular perches that provide entry to the feeding ports. To see busy feeders, each perch occupied by a hungry bird, underlines the importance of the food being provided. The filled feeders provide certainty, a reliable source of food at a challenging time of the year. This reliability of food provision has been explored in the 'wild', tested in the tough conditions of a Northern European forest during the depths of winter. Where birds have access to a reliable source of food, they have been shown to reduce the amount of time that they spend foraging; in turn, this lowers the risk of being predated because the birds can spend more time resting in thicker cover, rather than out in the open looking for food.

Something else to emerge from studies of wild birds in more natural circumstances is the finding that small birds tend to track a number of different feeding opportunities, regardless of how abundant favoured food is at any one of these. This makes sense, since tracking a range of food patches has clear survival benefits. If conditions change and one of the patches becomes unavailable, then the bird can quickly switch to another of those that it knows will be available nearby. This means that if your garden feeders happen to be left empty over a weekend, then the visiting birds are likely to know of several other gardens where similar feeding opportunities will be present. Knowing this doesn't seem to reduce the pressure that I feel when I see my nearly empty feeders; at such times there is an urge to top them up so that the birds don't go without. If this is a shared feeling, then it might go some way to explaining the £200 million that we spend collectively on bird food and other wild bird care products each year in the UK.

There are wider considerations to take into account when providing food for wild birds. In addition to thoughts about choosing the right foods, using the most appropriate feeders, and keeping things clean, there are also those about the origins of the food being provided and its impact on the birds that consume it. Have you ever thought about the origins of those sunflower seeds, or the peanuts that you provide? Looking at the packaging in my shed, I cannot tell you where the seeds and peanuts that fill my feeders have come from, at least not down to country, and certainly not down to the grower. Like all of the food products that we consume, bird food will have been grown somewhere, transported and, most likely, processed in some way.

Eastern Europe and Russia, which account for just under half of the worldwide production of sunflower seeds, are the most important source for our UK market, while your peanuts could have come from China, Argentina, India or the USA. Was the land they were grown on cleared for the purpose? Was growing the seed for bird food more lucrative than growing something else for human consumption? By feeding wild birds, am I contributing to climate change, damaging natural habitats or making things more difficult for people who have families to feed? While I aim to select food for my kitchen table that is organic and local, such choices become much more difficult when it comes to the sunflower seeds and peanuts destined for visiting birds. Maybe we should all ask more questions of the garden centres and mail order companies from whom we buy our bird food. And how environmentally friendly are their wider businesses?

Such considerations are important, and that we are thinking more about them is a positive sign, but they do seem a world away from the birds before me now, feeding just a few feet beyond my window in the crisp bright morning of this January day. To these birds the seeds and peanuts are a calorie-rich resource, providing some of what they will need to get through the day. Their behaviour, particularly the interactions between the different species, pulls my attention away from thoughts about consumer choice and global markets and back to the birds themselves. They are nature 'red in tooth and claw', individuals facing the challenges of finding food, shelter and, later in the year, a territory and a mate. They are partners, competitors, predators and prey. Theirs is a shorter existence than ours but equally packed with the needs that sustain life. Each one is an evolutionary marvel, shaped by its environment and through interaction with its fellow creatures. There is a joy that comes from watching them, from focusing in on this shared aspect of their lives, made possible through the food that I provide.

I let go of the guilt that comes from the awareness of my actions and simply delight in their presence. On this bright morning, my world is full of hope, beauty and joy, framed in the Christmas card simplicity of the garden bird scene before me.

* * *

As January nears its end, I take another delivery of bird food. Buying in bulk, I should have enough to see my garden birds through the remaining winter months. The metal bins in the shed are now full, the seed inside secure from the wood mice and bank voles that I know visit the shed at this time of the year. The remaining food, that which would not fit into the available bins, is balanced where I hope these small mammals will not notice it. If they do, then their sharp teeth will make quick work of the bags and spill the contents onto the floor below. It is a risk, but as long as it remains crisp and cold, the remaining food will be used within a week or so. The metal bins not only protect the food from the attentions of wintering rodents, but they also help to keep it dry and free from mould.

The bird feeders are filled daily now, the volume of food taken reflecting the larger numbers of birds visiting and the increased need for them to replenish fat reserves used to maintain body condition overnight. Once a fortnight the feeders are cleaned, washed and rinsed thoroughly and then left to air-dry in the shed. In their place is a second set of feeders, the two operated in rotation. There are times when I wonder if I should increase the number of feeders in use, but I reckon that the visiting birds have plenty of options locally, including just next door, so I am comfortable with the half dozen or so that are deployed around the garden. As well as the traditional seed feeders, filled with sunflower hearts – the staple food that I provide – there is a single feeder filled with mixed seed and another full of peanuts. The peanut feeder is a wire-mesh construction that allows birds to take pieces of peanut but not a whole nut. For now, the niger feeder remains in the shed, unused in many months because I found that the local goldfinches and siskins no longer visited it when they had the option of sunflower hearts in a feeder located close by. Niger is a small and expensive seed, favoured by birds with fine bills, but seemingly only when they are faced with strong competition for larger seeds. The mixed seed is also used on the bird table and on the ground, where it is taken by chaffinches and, occasionally, wintering brambling. A couple of suet blocks hang in their own wire-framed feeders nearby, and these are well used by visiting tits and starlings. I tend not to provide suet outside of the winter months, a personal preference but shaped by how quickly the blocks can develop mould under warm or damp conditions. If I had more starlings using the garden in the summer, then I suspect that the suet would be better used.

Pulling some logs from the covered wood pile, I notice that several from the top of the stack are covered with bird droppings. Looking up, I find more on the narrow brace that helps to support this homemade construction. A small bird, possibly a robin, wren or tit, has been using the woodstore as a winter roost, taking shelter each night from the worst of the weather. Intrigued, I check the remainder of the woodstore, but this appears to be the only bit of it being used by a roosting bird. This is the most sheltered end of the woodstore, the one tucked up against the workshop by the hedge, so perhaps this is why it has been favoured. The number of droppings suggests regular use and I make a mental note, reminding myself to make sure that I have enough logs in the house so that I do not have to come blundering around in the dark of evening and risk disturbing my visitor.

The end of January brings the RSPB Big Garden Birdwatch and garden birds gain prominence in newspaper and magazine articles, as well as across social media. For many families, the increased awareness of garden birds that results from the survey's promotion triggers renewed interest, perhaps leading to the purchase of a bird feeder and some food, or to participation in the survey itself. The survey, carried out over a single weekend, delivers a snapshot of our wintering bird community and introduces people to the idea of citizen science, where volunteers get to contribute to our understanding through structured scientific studies, accessible in their format and open to all. Surveys like this can reveal broad-scale patterns, such as differences in the structure of wintering bird communities across the country or between habitats. They can also reveal longer-term changes that might take place over time, such as those linked to a changing climate, even though the individual surveys may be strongly influenced by the vagaries of the weather during one particular winter. Some of the patterns seen in the Big Garden Birdwatch results reflect the changing fortunes of particular species, such as the decline in house sparrow and starling populations or the increase in those of woodpigeon. They can also reflect change in behaviour, such as the increasing tendency for blackcaps to winter in UK gardens. In such cases, it is the results of other citizen science projects or those carried out by academics that provide the supporting evidence or elucidate the underlying causes. Surveys like this help to change the way in which we look at our garden birds and place our own personal observations into a wider context. They add to our enjoyment, increase our understanding and, in many cases, help to shape our own behaviour and the ways in which we might better manage our gardens for birds and other wildlife.

Another daily job at this time of the year is the bird bath, which more often than not freezes solid overnight. Fortunately, this means that I can simply lift out the block of ice and fill the bath with fresh water. If it is particularly cold over a run of days, then a small pile of bird bath-shaped ice sculptures is slowly created over the course of the week. While filling the bird bath can seem like a never-ending job, the sight of its use by blackbirds, collared doves and robins reminds me of its value. The birds also have the option of visiting the wildlife pond. This can also freeze over at times, requiring me to open up a hole or two in the ice, but it copes with the cold conditions far better than the bird bath. The pond, which is very well used in the dry summer months, tends to be a lot quieter at this time of the year, a reflection of the changing needs of visiting birds.

February

I must admit to finding February something of a struggle; the optimism of the New Year has long faded and the reality of the remaining winter weeks, with their short days and poor weather, has always sat rather heavily upon me. While I appreciate the comfort of a warm fire and the company of family and a good book, I am at heart an outdoor person and I struggle with the lack of daylight, especially during the working week. To leave and return home in the dark, spending the day under the artificial glare of office lights, seems a poor way to spend the winter; better to be outside and to feel winter's bite. At least March offers glimpses of the coming spring. With February it is winter, like it or not. The working week can go by without delivering any sign of the birds who so mysteriously empty the feeders, feeders that are replenished in the dark of evening and which do not receive their first visitor until after I have left for the office come morning.

The ritual of filling the feeders reminds me of why, like so many others, I put out food for birds. I do this because I believe that it helps them, especially at this difficult time of year, and because I feel that many of the challenges that these birds face result from our actions. We have transformed the landscape, altering the balance of natural systems and replacing natural habitats with sprawling towns and acres of monoculture. Feeding the birds provides a tangible link between my life and theirs, reminding me that these are my neighbours, the creatures alongside which I live. I get to see who these neighbours are on those days when I am at home during daylight hours: the succession of thrushes, tits and finches that visit the bird table and feeders to take seed and suet. I know, from the scientific literature I have read and studied, that the provision of this food is not a straightforwardly beneficial act. I know that there are costs as well as benefits, both to the birds themselves and to the wider environment, and I know that in a perfect world the provision of such food would be unnecessary. But I also know that it can make a difference, that the food provided is taken on their terms, not ours, and that in the wider scheme of things it remains a small but not insignificant act. I have chosen to purchase and provide this resource, be it a sunflower heart, a peanut or some mixed seed, in the hope that it will be taken and used by a visiting bird. I may gain additional pleasure from watching the finch or tit take the seed, but the seed is just a resource and the bird does not see my part in its provision.

Whether we are providing food to wild birds for our benefit or for theirs, or indeed for both, this remains an activity that is widely practised in gardens across Britain. It is a cultural practice, present in many other countries, though not all. As we have noted already, the £200 million we spend each year in the UK on bird food and other bird care products, such as feeders, nest boxes and bird baths, provides a significant resource for wild birds. It has also been estimated that here in the UK there is at least one bird feeder for every nine potentially feeder-using birds, providing enough food to support 196 million individuals. Thanks to the data collected by 'citizen scientists' volunteering for the BTO, we have seen how the changing nature of this food provision has shaped the community of birds that visits and uses our gardens. Watching my feeders now, I am struck by the variety of birds that have come to take this food. In addition to house sparrows, tits and finches are blackbirds and woodpigeons, both able to take sunflower hearts from the seed feeder that hangs suspended from the fence next to the honeysuckle. While the blackbird can cope with the 'O'-shaped perches, the woodpigeon has to cling onto the honeysuckle before leaning across to take seed.

The feeders close to the house also provide an opportunity to watch the behaviour of the visiting birds and, in particular, the exchanges that take place between them. Central to many of these interactions are the dominance hierarchies that exist both between and within species. The visiting great tits, for example, are dominant over the smaller blue tits, which in turn are dominant over the coal tits, whose visits usually involve darting in to grab a seed that is then taken away to be eaten elsewhere. The same pattern can be seen within the finches, the larger greenfinch dominant over the smaller goldfinch and chaffinch. Within a species it is the adult males that take on the dominant roles, aggressively displacing adult females and younger birds of both sexes. As we shall see later in this book, social status is often signalled through some plumage characteristic, such as the male house sparrow's black bib, or the stripe that runs down a great tit's chest and onto its belly. Such signals are important because they help to reduce conflict and to identify a suitable mate.

February has started cold but dry, the conditions challenging enough to keep the bird feeders busy but not so challenging as to bring additional birds in from the wider countryside. It feels like business as usual at the bird feeders, the daily cycle of activity occasionally punctuated by a passing cat or the blur of a hunting sparrowhawk, sweeping in to make a kill or, more likely, to miss its intended target. Activity peaks soon after dawn, drops through the morning before a smaller lunchtime peak and then increases again just before dusk. The pattern of activity reflects the energetic requirements of the visiting birds. There is a need to replace spent energy reserves first thing, a small top-up early afternoon and the laying down of new reserves before retiring to roost just before dusk. Given a reliable supply of food, many small birds take on board the resources needed but not more, because this would reduce their flight performance and agility, making them more vulnerable to a predator.

There is also a pattern to how the different foods offered are used. The sunflower hearts are the favourite: oil-rich and with no husk to remove, they are easy to handle. From the coal tits through to woodpigeons, these seeds are a firm favourite. Although initially cultivated by Native Americans, and with its origins in North America, sunflower seed probably arrived in Europe through the Spanish ports. Its importance as a foodstuff, both for human consumption and the bird food market, owes much to Russian agronomists, who managed to increase the seed's edible oil content through selective breeding, lifting it from 20 per cent to nearly 50 per cent in just a few years. It was these highly edible oil lines that were reintroduced into North America, and from there into the European bird food market. Black sunflower seed, which used to be a staple of the bird food market, was introduced to the UK during the 1970s, but it is the sunflower hearts introduced in the early 2000s that have proved most popular, both with wild birds and with the people who feed them. Without any husk to remove, the birds can process the seeds more rapidly, and the lack of a husk also reduces the amount of waste that is dropped beneath the feeder, much to the relief of many of those who provide food for visiting birds. Over the same period, we have seen a decline in the provision and use of peanuts, which have a lower calorie content than sunflower seeds. Peanuts are still utilised by birds like great spotted woodpeckers and nuthatches, which can struggle to perch on standard feeders, and they are used by tits and finches when access to sunflower hearts is not possible – either because none are present or because the feeders containing them are being used by more dominant species or individuals.

A similar pattern can be seen with niger seed, which tends to be used where access to other, more favoured seeds is reduced. Niger seed, renamed and marketed as 'Nyjer' in the USA, in part to clarify its pronunciation and in part to avoid unfortunate associations with a similar-looking slang word, was initially found to be particularly favoured by goldfinches. Niger seed is very small, requiring a specially adapted feeder to prevent it from pouring out onto the ground. With its fine bill, goldfinch is one of only a few species able to handle the seed effectively, two others being siskin and lesser redpoll. There is a growing suggestion that niger is falling out of favour with goldfinches, which seem to prefer sunflower hearts, and this shift may be linked to a release from competition. As greenfinch populations declined following the emergence of the disease finch trichomonosis, so goldfinches may have had more opportunity to access sunflower hearts unmolested. With reduced levels of competition, the goldfinches are no longer forced to feed on less profitable alternatives. This suggestion is supported by observations that young goldfinches are more commonly seen on niger feeders than adults, again indicating the effects of dominance and competition. This also explains why my niger feeder now languishes in the depths of my shed.

Weather conditions can also influence the pattern of feeder use and the types of food taken. Suet-based products are well used during the colder winter months, though it is worth noting that they are also more popular at this time of the year with those who feed garden birds. This may be because of how these products are marketed, or it may be because they are more stable under cold conditions and hold their form better. Suet-based products, packed with calories and often easy to access, provide a quick energy boost and are available as blocks, balls or pellets, meaning they can be presented in different ways to best suit different species. Starlings, in particular, make good use of suet balls and blocks, often dominating the feeders holding these products to the exclusion of other birds.

One of the key things when providing food for wild birds is to match the food provided to the numbers and types of birds visiting. Presenting the food in a range of different ways, from hanging feeders, mesh feeders, the flat surface of a bird table, to the scattering of smaller seeds that I place on the ground and in cover, provides for a greater range of species. A wire-mesh feeder full of peanuts provides a good option for nuthatches or great spotted woodpeckers but it won't help chaffinches or dunnocks. These latter two species are most comfortable feeding on the ground, but both will use a bird table and, if it has a suitable perch, a standard hanging feeder. Food provision should also vary seasonally, matching the numbers of birds that arrive to feed, so that food isn't left on a table and uneaten at the end of the day. The numbers of birds visiting the garden is greater in the winter months, as local birds are joined by others arriving from the wider countryside or even overseas. Cold weather and snow on the Continent will push finches and thrushes farther west, boosting the numbers wintering here in the UK and almost certainly increasing the numbers visiting garden feeding stations. By watching the larger bird feeders, I can gauge how quickly the birds are getting through the food and respond accordingly. If this winter's promised snow arrives, then I can expect the demand to increase.

* * *

For once, the snow appears as forecast and brings with it new arrivals in the form of half a dozen redwings and a solitary fieldfare, which join the growing numbers of other birds to feed on the windfall apples that I have stored in the shed for just such an occasion. Both of these arrivals, new to the garden this year, are winter visitors to the UK, arriving in October and likely to remain here through into March or beyond. In the case of the fieldfare, with its grey head, grey rump and striking chestnut brown plumage, it is thought that up to a million individuals might make their way to our shores each winter. These birds are drawn mostly from breeding populations located in Norway and Sweden, but they may also include individuals from as far east as Finland or even Russia. The redwings will have come from a similar area; at least that is the case for the birds using my East Anglian garden. Redwings wintering in Ireland and the north of Britain are more likely to have come from the smaller Icelandic and Faroese populations.

Bigger than the blackbirds alongside which it is feeding, the fieldfare dominates the scene: full of attitude, it muscles in on a particularly fine windfall and begins to feed. This provides me with a chance to study the bird more closely through my binoculars, affording the opportunity to take in the subtle patterning of its breast and flanks, with their pale ochre wash and grey and black feather scaling. A line of soft white above the eye, coupled with a few dark feathers below, gives the bird something of a stern appearance, though this soon disappears as the fieldfare shifts its position to gain a better angle on the apple's soft pulp. The white of the lower chest and belly seems all the more intense because of the snow. The fieldfare is undoubtedly a very smart bird, which may be why it attracts the name *zorzal real* in Spain – the 'royal thrush'. The English name 'fieldfare' has more humble associations, being derived from the Anglo-Saxon *feld ware* – the 'traveller of the fields' – and alluding to both the bird's nomadic behaviour and habitat use. For much of the winter fieldfares feed on earthworms and insects taken from the fields, only venturing into gardens when frost or snow cover prevents access to the soil and its invertebrate residents.

Weather conditions also determine the movements and habitat use of the smaller redwings. Similar in size to a song thrush – so smaller than a blackbird – these dark little thrushes have a prominent pale cream stripe above the eye, and another that sweeps down from the base of the bill before curving up slightly as it contours back. The stripe above the eye is known to birders, and indeed other students of bird morphology, as the *supercilium* (literally 'above the eyelid') and is a useful identification feature, absent in our other common thrushes. The redwing derives its name from the rusty-red colouration of the underwing, which, extending onto the top of the flanks, can be seen below the folded wing. Less obvious than you might assume, especially in the poor winter light, it is still far more striking than the pale ochre wash present on the more familiar song thrush. Redwings are extremely susceptible to poor winter weather, particularly long periods of widespread frost or snow cover. Such conditions force the birds to move farther south and west in search of food, taking some individuals as far south as the Mediterranean. We sometimes see high levels of redwing mortality in those winters when very cold conditions extend across the Continent. Redwings are not faithful to their wintering sites, and an individual using a Norfolk garden in one winter could end up in Portugal or even Turkey the next.

One other interesting feature of both of these wintering thrushes is their inclusion on the UK's Birds of Conservation Concern Red List. Their presence on this list is due to the handful of breeding pairs of each species recorded annually in the UK. A dozen or more redwing pairs and fewer fieldfare appear in the annual report of the Rare Breeding Birds Panel, which collates this information from birdwatchers, BTO nest recorders and other individuals. The reports mostly refer to nesting attempts made in Scotland, reflecting the northerly distribution of the species across Europe, but occasional attempts are made farther south, including the pair of fieldfares that bred in Kent in 1991. To see these birds now, feeding alongside others in the snow, makes their arrival back in October seem all the more distant. Then, on a crisp but clear autumn evening, I could hear the calls of migrating redwings passing overhead in the darkness, the first of the winter's arrivals. Now, with just a few more weeks of cold weather ahead of me, it feels like the winter should soon be over and these wintering thrushes will begin their return journey northeast to breeding grounds located far from this small plot of mine.

The thrushes are not the only species to make use of the windfall apples but they do tend to dominate them; the robin gets an occasional look in, though spends far more time visiting the small dish of mealworms placed on the bird table. Although I do not get green woodpeckers in the garden here, I remember one snowy winter at my parents' house when one became a regular visitor to the windfall apples placed on the elevated flowerbed outside the kitchen window. The bird was a female, identifiable as such by the black stripe that ran from the base of the bill at an angle below the eye – males have a red stripe, edged with black. The bird had no difficulty in holding its own against the fieldfares and blackbirds, which quickly edged away whenever the woodpecker moved towards an apple. The woodpecker seemed somewhat nervous on the ground and was easily frightened by a movement within the kitchen, so we had to stand still and watch. Such a close encounter with this stunning bird was a real treat for a teenage birdwatcher.

Although the covering of snow is just a couple of inches deep, it is sufficient to restrict feeding opportunities for those smaller birds, like robins and dunnocks, that would normally feed among the leaf litter, looking for insects, spiders, earthworms and small seeds. For larger visitors, such as the pheasants, the snow presents less of a challenge and they simply dig down to reach the soil's surface below. The robins are alert to this behaviour and soon begin to follow a pheasant as it rakes over the borders. The response to this opportunity is similar to the approach that the robins adopt as they follow me around the garden when I am digging and weeding. In the wilder countryside within which they evolved, robins would have exhibited the same behaviour, following rooting wild boar, deer and wild cattle.

* * *

It is the low temperatures that cause problems for our smallest garden visitors, most notably wrens, goldcrests and long-tailed tits. These tiny birds – goldcrests weigh in at just five or six grams – suffer from the challenge that faces any small creature with a large surface-area-to-volume ratio, namely high rates of heat loss. At night, when these birds are unable to feed, they have to rely on fat reserves built up during the previous day if they are to maintain their body temperature at a safe level. With fewer invertebrates available, winter is already difficult for these insect-eating species and the long winter nights can be a particular challenge.

Some birds are able to lower their body temperature in a controlled manner, entering a state of hypothermia in order to manage the rate at which their fat reserves are used. This appears not to be the solution adopted by the goldcrest, which instead deploys a behavioural response, roosting communally to share body warmth and reduce heat loss. Studies have shown that the amount of energy that can be saved through entering hypothermia declines with decreasing ambient temperature and, importantly, also with decreasing body weight. This suggests that such a strategy would not be advantageous for a bird that is as small as a goldcrest, especially during the coldest nights of the year.

As a licensed bird ringer, I have been fortunate enough to handle goldcrests and to appreciate just how small and delicate these birds are. To look at one in the hand, with its large dark eyes and tiny wings, you wonder at its ability to survive the winter. Even more remarkably – and this is something that has been revealed thanks to the efforts of bird ringers working across Europe – many of the goldcrests wintering in Britain will have arrived from breeding grounds located much farther to the north and east, including from sites in Finland and Russia. That such a small bird can undertake such a long migration, or indeed winter so far north, is astonishing. During autumn, large numbers of goldcrests move south and west, with many crossing the North Sea to reach our shores. The goldcrest is one of the lightest birds in the world to undertake regular sea crossings, and it is little wonder that you often encounter exhausted individuals along the East Anglian coastline, newly arrived in October. In some years, and dependent upon the prevailing weather conditions, it is possible to see many hundreds of individuals recovering in the coastal scrub and dunes. On 11 October 1982, some 15,000 individuals were present on the Isle of May, off Scotland's east coast. Remarkably, while many goldcrests undertake these incredible migrational movements, others choose to remain on their northern breeding grounds. This leaves them exposed to conditions where it can be dark for as much as 18 hours each day, with overnight temperatures falling to −25°C. Even with communal roosting behaviour, these goldcrests may see their body weight fall by a fifth overnight, as they burn off fat reserves in an effort to stay warm.

Goldcrest roosts will typically feature two or more individuals huddled together within thick evergreen vegetation, their feathers fluffed up, tails pointing outwards and their bodies in contact with one another. This approach to communal roosting is repeated in the long-tailed tit, although in this case the roost tends to form along a branch, the birds roosting side by side in a line. An individual's position in the roost will be determined by its social status, the dominant individuals occupying the central positions and the less dominant birds forced to the outer margins. Under particularly cold conditions, it will be those individuals on the outside of the roost that will be most likely to succumb to the weather. Although long-tailed tit roosts may be formed in evergreen vegetation, such as ivy, they are more often encountered in dense and thorny shrub.

A slightly different approach is adopted by another small garden bird, the treecreeper. As its name suggests, the treecreeper specialises in feeding on insects and spiders taken from the bark of trees. The bird can sometimes be seen spiralling its way up from the base of a tree, moving up the trunk in a series of jerky, mouse-like movements. During the winter months, individual treecreepers may come together to roost in a cavity, much like wrens. However, early in the 1900s treecreepers were also observed to excavate their own roosting sites, creating shallow depressions in the soft bark of Wellingtonia trees that had been introduced into the parks of stately homes and arboretums. The behaviour has spread throughout the treecreeper population and has now been observed at sites across the UK.

Wrens usually select roost sites that are well insulated, such as a cavity in a tree or a nest box, and it is not uncommon to find several individuals sharing a site. There are occasional reports of winter roosts containing several dozen birds, the wrens arriving at the roost at dusk and leaving again at dawn, often over an extended period. These roosts may be used both within and between winters, and it is thought that the establishment of a roost may be initiated by one bird, whose vocalisations attract others. Within the roost itself the wrens huddle close together, and if sufficient birds are present then they may even roost several layers deep. Towards the end of the winter, the male whose territory contains the roost may become more selective in who he lets in, favouring females over males, but he does run the risk of being usurped by a persistent and more dominant individual.

Low temperatures can lead to the formation of ice on the bird bath or, if damp, on the branches and trunks of trees. This can be a problem for birds like the treecreeper, and to a lesser extent the nuthatch, which would normally forage over the bark for overwintering insects and spiders. With a layer of ice covering the trunk, these small invertebrates are safe from the treecreeper's probing bill. The freezing conditions mean that the bird bath needs closer attention; a floating ball helps to reduce ice build-up, but on these cold days the bird bath is still a solid block of ice come first light. This is easily removed, and fresh water added to the bath. Although I have seen it suggested that it is better to fill your bird bath with warm water, this makes little difference because the speed with which water cools has been found to vary at different water temperatures. It turns out that cold water is just fine.

Although the laying snow lingers for just a few days, the low temperatures leave the ground crisp and the bird feeders busy. It is during this part of the winter, especially if conditions are poor, that unexpected garden visitors may sometimes appear. A scattering of seed below the bird table may attract pheasants to a garden feeding station, though typically only if the garden in question is rural in nature; such gardens may also host other farmland species, including yellowhammers, reed buntings and tree sparrows. These seed-eating sparrows and buntings would have once relied on overwinter stubbles, with their spilt grain and bank of weed seeds. Because of changes to farming practices, these birds now turn to garden feeding stations where they can, something that is of particular importance in late February and March.

Cold weather can also deliver some truly unexpected visitors, though sadly not yet to my garden. Woodcock and water rail are perhaps the most striking and unexpected of these. The woodcock is a particularly unusual bird, with its long probing bill, huge eyes set high on its head, and beautifully patterned plumage.

It is a wader, but one that breeds in woodland and forages for worms and other invertebrates on farmland fields at night. Although much of its food is taken from the soil's surface, some is taken by the bird probing into soil with its sensitive bill. Little wonder then that cold winter weather can force woodcocks to forage more widely, very occasionally bringing these striking birds into gardens, where they may be seen turning over the leaf litter on flowerbeds in search of food. Winter weather conditions are behind the much more significant woodcock movements that see our breeding population of some 50,000 pairs supplemented by the arrival of birds from Scandinavia and Russia. Thanks to the warming effects of the Gulf Stream, Britain experiences relatively mild conditions in winter and this makes it an attractive destination for many birds, fleeing the tougher wintering conditions that dominate breeding areas located far to the north and east.

Another breeding resident whose population is swelled by the arrival of winter visitors from elsewhere is the water rail, a species that you would hardly expect to encounter in a garden. This secretive rail, which is related to the more familiar moorhen and coot, breeds in wetland habitats, from reed and sedge beds to smaller freshwater sites with dense vegetation. Water rails are omnivorous and feed on both plant and animal material, taking roots, shoots, insects, snails and small vertebrates, like newts, frogs and fish. Plant material is more important during the winter months, but when very cold weather limits access to favoured sites individuals may sometimes venture into gardens, either to forage around a garden pond or to feed under a bird table. Here, they have been seen to take small birds like finches and house sparrows, caught unawares by the rail's rapid strike. When the weather is tough, it pays to be versatile.

* * *

A shift in the weather systems and suddenly the cold conditions are gone, though now replaced by wet and windy weather. Out filling the feeders, I sense just how wet the ground has become. Full of moisture, it has more give now that the frost has released its grip on those top few inches. It feels as if we have turned a corner and that somewhere, some way up ahead, is the first sign of spring. As if to reinforce this thought, a mistle thrush kicks into song. I can't see the bird itself, but it must be in one of the taller trees several gardens over. Reminiscent of a blackbird's song, but sparser and more far-carrying, there is a hint of melancholy to the notes. I think this comes from the song's pacing – which feels slow and deliberate, lacking the warmth and exuberance that is a feature of the blackbird's song. It may also have something to do with the weather. This thrush seems to be throwing out a challenge to the winter, singing as it is in these blustery conditions, but it lacks conviction. Perhaps, like me, it can sense the change that has come with the arrival of warmer, wetter weather, but again, like me, it knows that winter is not done with us just yet.

The mistle thrush has little competition, save the robin's wistful notes, and its song becomes the backdrop to the following days, caught unexpectedly through an open window or heard when out on a walk. The mistle thrush is an early nester, the first eggs often laid before February is out, so its efforts now will be directed towards the establishment of a breeding territory. While some mistle thrushes are wide-ranging during the winter months, sometimes forming flocks with other thrushes, studies suggest that many are largely sedentary in their habits, remaining close to their breeding territory and perhaps even wintering there with their mate. The early nesting habit of this species, which may make two or three breeding attempts over the course of the season, is not without risks. These birds often nest high off the ground, favouring the base of a branch where it joins the trunk, and this can leave the nesting attempt vulnerable to windy conditions. Perhaps then, the song really is a challenge, flung out as it often is on windy days or in the teeth of a storm; little wonder that the mistle thrush is known locally as the 'storm cock'.

The mistle thrush is associated with mistletoe, both through its common name and the scientific name attributed to it by the nineteenth-century Swedish biologist Carl von Linné. The scientific name *Turdus viscivorus* means 'the thrush that devours mistletoe', and it originates from the Greek word used by Aristotle to describe this bird. Admittedly, Aristotle had been watching mistle thrushes feeding on a different species of mistletoe to ours; his, local to the Mediterranean, is *Viscum cruciatum* and has red berries. While we associate this thrush with our white-berried mistletoe, it actually shows a strong preference for the red berries of holly over those of mistletoe, even though it remains one of the key agents of dispersal for the latter species in the UK.

One afternoon, working in the garden and with the mistle thrush still in song nearby, a flock of starlings catches my eye. The flock, which numbers a few dozen birds, is one of several to have passed overhead within the last few minutes, all heading the same way. My curiosity aroused, I wander down the garden and away from the house to get a better view. It is then that I spot the much larger flock of starlings that has formed over the back of the fire station and off towards a distant housing estate. There are hundreds and hundreds of birds in the flock, with more joining all the time, and it is quickly evident that this is a pre-roost murmuration. Starling murmurations are a true wildlife spectacle, a gathering of avian life bettered in the UK only by the huge winter flocks of waders that come together at a few favoured high-tide roost sites around our coast.

From a distance, the murmuration appears as a smudge on the horizon, though moving in a way that quickly distinguishes it from a rain-bearing cloud or the smoke from a chimney. On closer viewing the smudge resolves itself to become a multitude of small black shapes, each one a starling and each a part of this vast aerial ballet. That's the thing about a murmuration: this is not a flock in some disjointed sense, some loose gathering of independent individuals. Instead it takes on a life of its own, moving and pulsating in ways that suggest this is a single organism rather than a multitude of individuals. Waves of shadow pulse across the murmuration as individual starlings turn in unison, each one responding to the movements of the birds around it. The presence of this shifting, pulsating mass attracts other flocks, swelling the ranks of birds that will soon drop into the roost site above which the starlings are circling. In this case the roost is a line of mature conifers, densely planted.

The presence of the starlings has not gone unnoticed and I watch as a larger shape, one of the local sparrowhawks, sweeps down and punches a hole in the mass of birds. It misses, perhaps confused by the numbers of starlings present and unable to pick out a target. There is safety in numbers, as this coming together demonstrates. The ballet continues as the light begins to go, the glow of the sky now lit from below with street lighting and the headlights of passing cars. Then, at some hidden signal, the vast flock begins to move on a narrowing axis, centred on the conifers. The ripple across the flock flexes and bulges towards the trees, this action repeated again and again over several minutes. Suddenly, part of the mass breaks away and there is an audible rushing sound as a section of the flock drops into the trees like a flight of English arrows finding their mark at Agincourt. The process repeats itself, with more and more birds dropping into the roost to leave a shrinking cloud in the sky above. I turn away, conscious that the garden is now rather dark and that I have tools to find and put away. Later in the evening, I slip out and wander up the road over which the dark conifers loom. There is a soft chattering from the trees, the voices of a thousand starlings settling to roost and reminiscent of a theatre audience in the moments before the lights go down and the curtain lifts.

Communal roosting of this kind appears to have several functions. As well as providing safety in numbers, the roost delivers warmth and shelter. Just as with the goldcrests and long-tailed tits, there is a hierarchy to the roost: those positions located towards the centre are the best, and these are the ones occupied by the most dominant individuals. Communal roosts may also allow individual birds to assess the condition of their neighbours, enabling them to identify individuals who appear to have had a good day's feeding; these are the ones to follow when the roost begins to break up the following morning. Departing birds leave the roost in smaller numbers come morning, and there is none of the spectacle of the previous evening. The roost itself may be used night after night, though it is likely to be abandoned if it attracts too much attention from sparrowhawks or human residents, the latter unhappy to see their cars and patios splattered with droppings.

Starlings are not the only species to deliver a pre-roost spectacle. Not far from here, in mid-Norfolk, is a mixed roost of rooks and jackdaws, numbering several hundred birds that congregate each evening to roost in a line of trees alongside a cemetery. For part of the winter, and dependent on the relative times of dusk and dawn, I either pass them coming into roost or departing from it as I make my daily commute to and from work. There is something impressive about seeing even this number of birds together in such a small space, but this roost is insignificant when compared to the much larger mixed corvid roost that forms on the other side of the county at Buckenham Carrs, an extensive area of woodland located in the broad valley of the River Yare. Here, many thousands of rooks and jackdaws come together to roost, dominating the winter skyline as they gather. The spectacle has all the power that you would expect: the vast numbers of black birds, the fading light, the silhouetted trees, the cacophony of calls, and the rich weight of folklore associated with these birds. Knowledge of the roost has permeated far and wide, thanks in part to Mark Cocker's wonderfully evocative description of the roost in his book *Crow Country*.

A few days later, I visit one of the supermarkets near work. Walking across the car park towards the store, I notice a characteristic white splashing that peppers the concrete slabs marking the path between the rows of cars to the store's big glass doors. The splashing is not entirely random; each of the small London plane trees planted at intervals along the path has a circle of white splashing around it, extending out almost towards the edge of the tree's bare canopy. This is evidence of another communal roost, a nightly gathering of pied wagtails that winter here and presumably gain some warmth from the car park's lights that serve human visitors to this 24-hour retail opportunity. There are one or two wagtails already here, as I can hear their cheerful 'zlee-vitt' calls, but it will not be until later in the afternoon that dozens more will arrive to join the roost. I know of a few such sites locally; one in a nearby market town hosts the birds in the town centre, just outside Britain's smallest public house. Here they can be seen sat amid the Christmas lights, although the late-night shoppers are almost certainly oblivious to their presence. Pied wagtail is a bird I have never had in any of the gardens I have owned, even though they have always been on the garden list thanks to birds seen or heard flying over. They prefer gardens that are quite open, with a sufficient area of lawn over which these active birds can pace, tail pumping up and down, in search of small flies and other insect prey.

As the final few days of this short month are reached, so the thermometer lifts and it feels like a change is on the way. A couple of new feeders are added to the garden, replacing one that has finally given up the ghost. Both of the new feeders have metal perches and a metal base and lid, which should reduce the chances of squirrel damage. Importantly, both can be taken apart with ease to allow good access for cleaning, a job I try to do every other week. The act of adding the new feeders feels as if it is contributing to the coming change of season which, with luck, March will bring.

March

March has come in like a lion, the wind and rain buffeting the garden and shaking seed from the hanging feeders. It is a good day to be inside: a few odd jobs, then tea and a book by the fire. Outside, a succession of birds make quick visits to the feeders, the need to find food keeping them active. Under the shelter of the hedge, a blackbird turns over last year's leaves. Every now and then one of the leaves gets caught by the wind and spirals up into the air, coming down to land on the lawn just a few feet away. It feels like winter will be slow to release her grip this year, despite those few late February days that hinted at the approaching spring, and I am left wondering whether March will leave like a lamb, as the old proverb runs.

A few days later and the wind has returned, though thankfully this time without the rain. The local jackdaws have left their chimney-pot haunts and taken to the sky, making the most of the bright and blustery conditions. They seem almost playful in the wind, riding each gust and then checking back against it like skilled yacht-owners effecting a quick turn. There must be two dozen birds in the air together, not some ragged flock but a neat arrangement of the colony's pairs. Each pair of birds flies together, a few feet apart, but close enough so that it quickly becomes obvious to the human observer which are the established pairs. Jackdaw society is based on hierarchy, the species nesting semi-colonially, and an individual's social status typically increases with its age. High-ranked jackdaws have a shorter lifespan, so maintaining such a lofty position within the colony must come at a price. I wonder which of these birds, barrelling on the wind, holds pride of place in this little community.

Jackdaws usually pair for life and remain with each other throughout the year. Both sexes invest in the pair bond through a series of characteristic behaviours. These involve 'allopreening', in which an individual will gently preen its mate, and courtship feeding, where the male offers food to the female. Interestingly, courtship feeding in jackdaws appears to be a year-round behaviour. One of the reasons for such devotion is probably the challenges that jackdaw pairs face in trying to rear their brood of four or five chicks. Feeding conditions during the chick-rearing stage can often prove difficult, and both parents have to work hard to find sufficient food. These birds use a ploy rarely seen in other garden birds in an attempt to further increase their chances of rearing at least some young during the breeding season, a breeding season that will begin in earnest in just a few weeks' time. While most garden birds begin the incubation of their clutch only when the last or penultimate egg has been laid, jackdaws initiate incubation from the second or third egg. A consequence of this is that the eggs do not all hatch at the same time, but instead do so over a number of days. This creates a size hierarchy within the brood, something that favours the largest chicks when food is hard to find because they will always outcompete their smaller siblings. If conditions are poor then the smallest chick will be the first to die, the reduction in the number of mouths to feed increasing the survival chances of the remaining individuals. If all the chicks were of a similar age, and therefore size, then the chances are that tough times would lead to the loss of the entire brood. This behaviour is more commonly seen in birds of prey and owls, where the unpredictability of feeding conditions can be a particular challenge.

Another odd feature is the tendency for those jackdaw broods produced late in the breeding season to have a female bias, both at hatching and at fledging. Male chicks, it seems, experience a higher mortality rate than female chicks, so investing in female chicks makes more sense, especially towards the end of the season when feeding conditions become even more challenging. Quite how female jackdaws manage to skew the sex ratio of their clutch of eggs is an interesting puzzle, and one that is still to be answered. Watching these jackdaws now, mastering the windy conditions in an almost joyful manner, makes me wonder how they will fare come May and June, when there will be broods of hungry young to feed.

For other garden birds, the breeding season is already underway. Evergreen cover, such as that provided by ivy, ornamental conifers and some hedging, provides the first nesting opportunities of the year, and it is here that the earliest blackbird and song thrush nests are likely to be found. With the rest of the garden looking rather sparse, and the coming season's leaves still tightly held within their buds, this cover also attracts roosting birds and the noisy social gatherings of house sparrows. Later in the year, it will hold nesting dunnocks and wrens, perhaps even the soft mossy cup of a greenfinch, beautifully crafted and lined with feathers.

Although ivy is sometimes viewed as a hooligan, thought by some to be responsible for damaging masonry and killing the trees and shrubs over which it climbs, it is a terrific plant for wildlife. In addition to the nesting cover and shelter that it provides, ivy also delivers a source of nectar and pollen for late-season insects. Here in the UK, ivy flowers from late August to late November. Its fruits – black, almost berry-like, and held in tight little clusters – provide food for birds during the second part of the winter when few other berries are available. In a hard winter, thrushes and woodpigeons may strip most of the berries during January and February, but in most years some of the crop remains through into April or even May. The berries, which are often eaten by woodpigeons before they are fully ripe, have a high fat content compared to other bird-dispersed fruits.

Mature ivy, perhaps established on an old garden wall or around the trunk of a tree, is worth checking for nests from late March. Evidence that your local blackbirds and song thrushes have started to build is initially provided by the sight of a female collecting nesting material. In thrushes and blackbirds, it is the female who builds the nest and, with the exception of mistle thrush, has sole responsibility for incubating the eggs. Thrush and blackbird nests are bulky affairs, constructed from moss, grass and plant stems; surprisingly, they also incorporate a layer of mud into the nest's construction. The birds start the nest with a foundation of moss, and it is onto this that the grass and plant material used for the main body of the structure is then added. A female blackbird will incorporate a layer of plastered mud into the nest before the addition of the final layer of fine material that will cup the clutch of eggs. A similar approach is adopted by mistle thrush, whose nests are usually placed higher off the ground in tall shrubs and trees.

The song thrush takes a slightly different approach. Instead of the rough and ready mud layer seen in the nests of its close relatives, the song thrush delivers a smooth and unlined nest cup made from fragments of rotten wood and mud, sometimes with the addition of dung, all cemented together with saliva. Once dry, the nest cup has the appearance of chipboard, smooth to the touch, neat in finish, and forming a pale grey background against which the song thrush's bright blue eggs stand out. One might imagine that the hard lining would be a risky substrate for something as delicate as a song thrush egg, especially when compared to the soft lining characteristic of virtually every other songbird's nest, but it seems to work. There has been a fair bit of speculation as to why the song thrush favours this approach to lining its nest. Research has demonstrated that the chipboard cup reduces heat loss from the nest during incubation, but this is something that blackbirds also achieve through the layer of mud incorporated into their nest, though to a lesser degree. Some researchers suggest that the smooth lining might reduce opportunities for nest parasites, such as fleas, mites and lice, to go unnoticed, but this hypothesis has yet to be formally tested. Whatever the reason, the smooth lining and bright blue eggs make this nest easy to identify.

These are not the only nesting attempts that may be underway in the garden during March. Long-tailed tits, which may have started nest building during the final days of February, often have a completed clutch of eggs by the end of March. These delicate little birds construct a beautiful domed nest from moss, bound together with silk stolen from spider webs and decorated with grey-coloured lichen. The nest is built from the bottom up over many days and, once constructed, has to be lined with large quantities of small feathers, scavenged from the environment. During its initial stages, when it still has the form of a simple cup, it is easily mistaken for a chaffinch nest, though this species rarely nests so early in the year.

Studies looking at nest construction in the long-tailed tit have recorded up to 2,600 feathers in a single nest, the feather lining contributing up to 40 per cent of the nest's mass. Those long-tailed tit nests built early in the year contain more feathers than those constructed later, suggesting that the insulative properties of the feathers are important for incubation and that long-tailed tits can adjust the quantity of feathers used to match the ambient temperature outside of the nest. Having said this, it is worth noting that many of the nests constructed later in the season are re-nesting attempts made by pairs whose initial attempt failed because of predation, and these birds may be under greater time pressure to get on with their breeding attempt, leading them to cut corners during the construction phase.

Long-tailed tits are pretty relaxed when nest building, often noisy and certainly fairly obvious when carrying feathers to the bush or shrub in which the nest has been placed. This, coupled with the lack of leaf cover so early in the season, may be a reason why the nests are prone to predation by carrion crows and magpies. Many a nest will be broken open by a carrion crow in the hope of finding eggs, though many predation events find only an empty nest, the first egg yet to be laid. The effort that must go into constructing this complex nest is staggering and it is little wonder that both members of the pair are involved in its construction. As we've just noted, for those pairs building a new nest after the first has been lost there may be an additional urgency to get the job done. Not only do second nests contain fewer feathers, but they are also built more rapidly. Long-tailed tits often place their nests in a thorny bush or shrub, favouring bramble, hawthorn, blackthorn or garden favourites like *Berberis*. Some pairs use the thicker cover provided by ivy or *Clematis*, while others rely on the camouflage provided by a covering of lichen, for their nests placed in the fork of a tree.

* * *

Away from the nesting cover, the garden feeding station is still busy with activity. One of the features of the feeding station during these late winter days is the arrival of siskins. This delicate finch, with its relatively small head and short tail, could easily be overlooked, its green and yellow tones similar to those of the greenfinch. The dark wings, with their prominent wing bar, coupled with the male's black crown and bib are useful features to look out for. The siskin's bill is also much finer than that of the greenfinch, reflecting a diet that is dominated by the small seeds of alder, birch, spruce and pine. The greenfinch favours larger seeds and so requires a stouter bill to deal with them. Despite their small size, siskins can be surprisingly aggressive at hanging feeders, often noisy and unafraid to have a go at larger, more dominant species. They usually arrive in a small flock, the bickering calls alerting me to their presence long before they have dropped down to the feeders from the taller trees that surround the garden.

Siskins tend to visit gardens that are located near to areas of coniferous woodland but they can range widely during the winter months, especially if wider countryside crops of favoured tree seeds have been poor. Up until the 1950s, the UK siskin population was largely restricted to the Scottish Highlands, where the birds nested high in the tops of pine trees. During the winter months, the birds would move to lower ground, though invariably still remaining within forested areas. The expansion of commercial forestry operations elsewhere in the UK enabled the breeding population to expand, increasing the siskin's breeding range, and from here it seems to have been a small step to move into gardens to exploit the peanuts and seeds on offer at feeding stations. It has been suggested that the initial move into gardens was facilitated by the practice of providing peanuts in red mesh bags, which some authors have suggested resembled giant alder cones, but there is only anecdotal evidence to support this view. Today, siskins are attracted to sunflower hearts, peanuts and niger seed, and they remain one of only a few species to be able to handle the tiny niger seeds that larger-billed birds rarely waste their time on.

One of the most interesting aspects of feeding station use by siskins is the extent to which this can vary from one year to the next. This changing pattern is linked to the size of the tree seed crop available within the wider countryside, which itself can change dramatically between years. Fluctuation in the size of the seed crop is a natural phenomenon, recognised in many different tree species, and part of a strategy to increase the chances of seeds escaping the unwelcome attentions of seed predators like birds and small mammals. If a tree were to invest a lot of energy in producing large quantities of seeds year after year, then this would lead to an increase in the populations of its seed predators – which would benefit from the abundance of seed available during the autumn and winter months. With an increased population of seed predators, the chances of the tree's seed surviving through to germination would be reduced. Instead, a tree that produces a lot of seed in one autumn will produce little the following year, something that acts as a check on seed predator populations, which suddenly find that there isn't enough seed to go around. This approach only works if all of the trees in an area produce a bumper crop at the same time, something that is known as a masting event. The production of a bumper crop overwhelms the seed predators – there is simply more seed than the mice, voles and birds know what to do with – meaning that a greater proportion of the seeds will go on to germinate and to develop into saplings.

All this is important because it drives the use of garden feeding stations by siskins, and indeed by other seed-eating species like nuthatches, great tits, jays and chaffinches. We can see the influence of these masting events in the observations generated by participants in the BTO's weekly Garden BirdWatch project, which has been running since 1995, and the longer-running Garden Bird Feeding Survey, also operated by BTO, in this case since 1970. Analysis of the Garden BirdWatch data alongside a measure of the size of the sitka spruce seed crop in the wider countryside – the tree seed favoured by siskins – neatly demonstrates the link between the two. Garden feeder use is greater in those years when the sitka spruce crop is poor but drops dramatically during mast years. A similar analysis of Garden Bird Feeding Survey data found a matching pattern for several other garden birds, though this time the relationship was with several broadleaf tree crops, including oak and beech. Just as interesting is the tendency for feeding station use by siskins to be greater on damp days than dry ones. It is known that sitka spruce cones open up on dry days to release their seed but remain firmly closed when the conditions are damp. This leaves the seed available to siskins and other seed-eaters on dry days but not damp ones, which is why garden feeding stations become more important on wet or overcast days in late winter.

The extent to which different bird species use our bird tables and hanging feeders is in part determined by natural food supplies elsewhere, as ably demonstrated by the work on siskins just mentioned. Birds are mobile and can easily move between areas in order to find food. Such movements underpin the long-distance migration of birds to and from Africa. In this case, the movement north into Europe enables insect-eating species to take advantage of the long summer days that we experience and the associated seasonal boom in insect populations. During the autumn, as these insect populations begin to wane, the migrants retreat south. The influx of seed-eating finches and berry-eating thrushes into the UK during autumn reflects similar migrational movements, made to exploit a seasonal abundance of food. Even smaller-scale movements may bring seed-eating yellow-hammers and reed buntings into rural gardens in the winter, as we will see later in the chapter, with individual birds moving between habitats (from farmland to gardens) to take advantage of the food on offer.

Another factor in the extent to which individual species use garden feeding stations is diet. Take a look at the community of birds visiting your hanging feeders and you will see that it is dominated by seed-eating species, with relatively few insectivorous birds in evidence. Wrens, long-tailed tits and goldcrests are three of the most familiar of these insect-eating birds, and all three are less common in gardens than the seed-eating finches and sparrows. Other (largely) insectivorous species, such as blackbirds and starlings, are more common in gardens because they may take other foodstuffs or because they feed on the soil-dwelling invertebrates found within our lawns and flowerbeds. For such species it is other feeding opportunities within the garden that are more important; this is something to which we will return later in the year.

* * *

Late March can be an interesting time at a garden feeding station, not least because it is when we start to see a crossover between the winter community of visiting birds and the one that will use the garden through spring and on into the breeding season. We've already seen how the community of finches and thrushes using the bird table, windfall apples and hanging feeders is made up of birds drawn from a spread of different breeding populations. In with the chaffinches and blackbirds that will go on to breed locally are other individuals of the same species, whose breeding attempts will be made in Germany, Poland, Norway or even Russia. These visitors go largely unnoticed, being similar in appearance to those individuals that will perhaps spend their whole life within just a few miles of your garden.

Watch your own garden regularly throughout the year and you will notice the increasing numbers of blackbirds and chaffinches that appear during the winter, clearly indicating some form of movement into gardens; most people assume that these are birds from the surrounding countryside rather than from farther afield. While some of these individuals will be from the surrounding countryside, come in search of easier feeding opportunities, the presence of individuals from overseas has been proven, thanks to the efforts of volunteer bird ringers, trained and licensed to catch wild birds and fit them with uniquely numbered rings. Blackbirds caught and ringed in UK gardens during winter have been found to have come from, or to later return to, breeding areas located from eastern France, north to Denmark, Sweden and Norway, and east to Poland, Finland and Russia. With some 26,000 blackbirds ringed each year across Britain and Ireland, we now have a good understanding of the scale of movement into the UK from overseas. There have, for example, been nearly 8,000 recorded movements of ringed blackbirds between Britain and Ireland and elsewhere, including nearly 1,800 exchanges with Norway and over 1,300 exchanges with Germany. It must be remembered that this pattern of movements is not static; it may change from year to year, in relation to food availability, or over the longer term because of a changing climate. For example, evidence from bird ringing shows that many blackbirds from Scotland and northern England used to migrate to Ireland for the winter, but this behaviour has been largely abandoned because of our increasingly mild winters, a consequence of the changing climate.

By March, as we have already noted, some of our resident blackbirds will have already initiated their first breeding attempts of the year. Many of our resident males will have developed the bright crocus-yellow eye ring and bill characteristic of their breeding plumage, a colour trait that shows a strong seasonal pattern. Look at the blackbirds feeding on your lawn in March, and any that display the crocus-yellow eye ring and bill will be local birds, while those that have the all-black plumage of an adult male but which lack the bright yellow bill are likely to be winter visitors. Bill colour in male blackbirds has been shown to play a role in mate selection, with females preferring males with brighter bills; the first nesting attempts of the year invariably involve those males with the brightest bills. The crocus-yellow colour of the bill is produced by carotenoid pigments, which come from the bird's diet and are difficult to acquire. Because they may reflect individual foraging abilities, this carotenoid-based bill colouration is thought to provide an honest signal of a male's quality as a father. Take a closer look at your blackbirds and you'll start to recognise these subtle but important differences.

Other winter visitors are more obvious. Bramblings, redwings and fieldfares do not breed here – at least not in any meaningful numbers – and so they stand out as winter visitors, marking the changing seasons through their arrival and departure. By late March, many of these visiting birds will be preparing to move back to their breeding grounds, although the movements themselves typically do not begin until April. Some will just be beginning to acquire their breeding plumage. In bramblings, for example, the male's striking black breeding plumage – which extends across the head and down onto the back, and which contrasts so beautifully with the warm orange breast and white underparts – is rarely seen in the UK. During the winter months, the male brambling's breeding plumage is hidden by buff brown feather tips, but by late March these have begun to wear away and you get a hint of what will soon be revealed. In some years, when a late start to spring can keep our wintering bramblings here longer than usual, you may be fortunate enough to see a male with well-developed plumage, his head far darker than that usually seen. The same process occurs in chaffinches, the dull winter crown plumage gradually worn away from late winter through into spring, revealing the steel grey finery that will eventually replace it.

The reed bunting is another species in which the male's breeding plumage is acquired through feather wear rather than feather replacement, in this case the resulting transformation delivering a striking black head and black bib. The reed bunting is an uncommon garden visitor, most likely to turn up in those rural gardens located within arable farmland or close to wetlands and other damp habitats. Its presence at garden feeding stations peaks in March, a pattern also seen in two other farmland birds – yellowhammer and tree sparrow. All three species feed on seeds, including spilt grain, and in the past these birds would have made use of overwinter stubbles. With the change in farming practices, and the move to autumn-sowing of cereal crops, overwinter stubble has been largely lost from the farmed landscape. This has almost certainly removed an important source of food for these 'granivorous' (seed-eating) birds and it probably contributed to the long-term declines witnessed in their breeding populations. In late winter, with farmland seed supplies at their lowest ebb, reed buntings, yellowhammers and tree sparrows have had to turn to garden feeding stations, hence the seasonal peak evident in garden bird studies like BTO's Garden BirdWatch and the Garden Bird Feeding Survey.

* * *

Another farmland bird that shows a peak in garden use in late winter and early spring is the rook, a species that is still persecuted in the wider countryside because of its preference for cereals. While wary of humans more generally, those rooks visiting garden feeding stations can be surprisingly confiding and approachable. In fact, they can sometimes become something of a nuisance, monopolising feeders and digging holes in the lawn. For several years, I lived next door to a small rookery and was able to listen in on their lives. I say 'listen' rather than 'watch' because of the volume and diversity of calls that emanate from an active rookery and the challenge of watching activity taking place many metres above your head. By late March, most rookeries will have nests containing a complete clutch of eggs, and by this stage the volume of activity drops a little. Over the previous weeks, there will have been much coming and going, the birds collecting suitable twigs to add to existing nests or with which to construct new ones. In many cases, these twigs will be taken from the ground or broken off from living trees – the ancient walnut in my old garden used to take a bit of battering – but rooks are not averse to a bit of thievery and will raid unguarded nests to source material. To prevent her nest being targeted, a female will often remain on guard while her mate collects new nesting material from elsewhere, only joining him in this venture once the pair has started to line the nest with grasses, rootlets and moss.

Most English rookeries are small, containing fewer than fifty nests, and many may number just a dozen or so active nests. In part this reflects the levels of persecution that rooks face; many nests are 'shot out' by landowners keen to reduce damage to cereal crops and this leads to birds dispersing away from targeted colonies to breed elsewhere. But colony size will also be limited by the availability of food within the local area and, indeed, of suitable trees in which a rookery can be established. There are perhaps only a few English rookeries left with more than 300 nests and certainly not the four-figure counts that have been noted in the past.

If you are fortunate enough to have rooks visiting your garden, then it is possible to observe both the iridescent beauty of their plumage and the level of their intelligence. Faced with a problem, such as how to access a feeder full of seed that is hanging from a branch on a length of twine, a rook can be surprisingly quick to come up with a solution. Individual rooks have, for example, been seen to pull up the twine attached to a hanging feeder with their beak, securely clasping the length with their foot before reaching down to pull up another section. After repeating this trick a few times, the rook will have hauled the feeder to within reach and it can then feed at leisure. A similar approach may be used with a feeder full of suet balls, the lid then removed and the balls tipped to the ground, where they are then quickly demolished. Despite the use of garden feeders, the rook is essentially a bird of open grassland, where it uses its strong bill to probe for soil-dwelling invertebrates like earthworms and beetle larvae. A garden lawn can prove just as attractive, and it is quite a sight to see a flock of rooks working their way across the grass with their confident upright posture and careful probing movements.

The rook is not the only member of the crow family to make use of gardens and garden feeding stations. Magpies and jackdaws are regular visitors to garden feeders, while jays are reported from about one in ten gardens and carrion crows from one in three. All four of these species are more catholic in their dietary choices than the rook, something that has enabled them to make greater use of urban and suburban sites. Such is the opportunistic nature of the magpie, for example, that it has taken readily to the built environment and can be found nesting in urban parks, in the taller roadside scrub that grows alongside bypasses, and in some larger urban gardens. Although the magpie diet is dominated by invertebrates, especially insects, it also takes fruit, seeds, berries, carrion, small mammals and the eggs and young of other birds. Carrion, in the form of unfortunate birds and mammals killed through collision with motor vehicles, seems to be an important source of protein and, along with foxes and carrion crows, magpies play an important role in carcass removal and recycling.

With reduced levels of persecution and an abundance of food, it is hardly surprising that magpies find the urban environment such a welcoming one. As with the rook, the first magpie eggs may be laid in February, with the core breeding season running from March through to mid-June. The magpie's substantial nest, which is unusual in being topped with a loose canopy of twigs, is easily spotted during the late winter months before the leaves emerge. It looks like an oversized football in silhouette, dark in the bottom half but more open above because of the loose nature of the twig canopy. Not all magpie nests have a canopy; research has shown that it is the older, more established birds that are most likely to add one to their nest, while young and inexperienced birds often leave their nest uncovered. Pairs often re-use the nest from a previous season or build a new one nearby, making it fairly easy to find the nest site of your local pair. Individuals can be regular visitors to a bird table, especially if you offer kitchen scraps or suet. While much of this food will be consumed immediately, the magpie is one of a number of species to carry food items off to be hoarded elsewhere for later consumption; this is something that we will look at in more detail come autumn.

Sit watching a visiting magpie and you get an idea that it is an active forager, taking insects and other invertebrates as it walks confidently over areas of short grass. Lawns and other forms of short vegetation appear to be more important for urban magpies than is the case for their rural cousins. Certainly, urban magpies face different challenges to those living in the wider countryside. While urban birds benefit from reduced levels of persecution and show significantly better breeding success, they produce fewer chicks per nesting attempt. This suggests that food supply, or rather the supply of invertebrates for growing magpie chicks, is limited within gardens and across the wider urban environment. This is a pattern that is also evident in many other 'garden' bird species. Despite the broad range of foodstuffs taken by adult magpies, insects are of particular importance and the diet of urban chicks remains dominated by invertebrate prey, especially beetles, a pattern also evident within the wider countryside. Urban magpies nest earlier in the year than their rural counterparts, almost certainly a reflection of the supplementary food they receive at garden feeding stations, and they also tend to nest lower down. Whether this latter observation reflects choice on the part of urban birds – perhaps able to change nest location because of reduced rates of nest predation – is unclear, and it may simply be that there are fewer suitable taller trees and shrubs in our towns and cities than is the case elsewhere. One pattern that is consistent across habitats, however, is selection for breeding sites with greater levels of leafy cover; clearly you do not want your nest to be too obvious, regardless of the protection that adding a canopy might afford.

* * *

The magpie is not the only black-and-white bird to have visited our lawn in recent days. There is also a male blackbird whose plumage carries a striking set of white feathers; this unusual plumage abnormality gives the bird something of a piebald appearance. While I have sometimes encountered other blackbirds, house sparrows, carrion crows and jackdaws with the odd white feather in their otherwise normal plumage, this is the first individual that I have seen with such a significant abnormality. The presence of white feathers in this blackbird's plumage results from a lack of the melanin pigment responsible for the bird's more usual colouration.

Special cells deliver pigment to a growing feather, a process driven by an enzyme present within the cells but which is also dependent upon amino acids obtained in the bird's diet. Any disruption to this process, whether caused by a temporary external factor – such as a dietary problem – or some mutation that the bird has inherited, can alter the pigmentation of the resulting feathers. While we will all be familiar with one such aberration, albinism, which results in individuals with all white plumage, pink eyes and pink legs, there are others. The presence of some white feathers within other correctly coloured plumage is the likely consequence of either 'leucism' or 'progressive greying', two other forms of aberration not uncommon in birds. A feature of leucism is that the pattern of white feathering is typically patchy in nature, most commonly affecting the head, belly and wing-tips. It is often symmetrical, such that matching feathers on each side of the bird both lack pigment. This is not the case for the male blackbird visiting our garden, whose pattern of white feathering is somewhat random and, therefore, suggestive of 'progressive greying'.

The blackbird continues to visit for several days but then disappears; perhaps it was passing through en route to a breeding territory located somewhere within the wider countryside, or perhaps it was predated by one of the many cats that prowl the local gardens. Either way, its presence over these last few days has reinforced the sense that each of the birds visiting the garden, bird table and hanging feeders is an individual, with its own life story. While we can recognise different plumages – male and female, adult and juvenile – for many of the garden birds that visit, making the step to recognising individuals is much more difficult, and only rarely do we have a chance to get to watch and know a bird as an individual. There are features, such as a plumage abnormality or some unusual behaviour, that may serve to mark out a particular bird, and the more time you spend watching the better your chances of making these scarce connections.

As March nears its end, the garden begins to offer more glimpses of the coming spring, both in terms of the emerging flowers and the growing number of birds contributing to the developing dawn chorus. To the rich tones of the blackbird, the thrice-repeated notes of the song thrush and delicate warbling tune of the dunnock is added the regular 'teacher teacher' motif of the great tit, a species whose first nesting attempts are still some weeks away. There is a sense that winter has reached its end and that the new season is about to burst forth. With the passing of the spring equinox this sense is even more apparent, spurred on by a run of still days, bright and touched with a growing warmth. This year, March truly has entered like a lion and left like a lamb, a lamb bathed in the light from lengthening days and all of the hope that this brings forth.

April

April's arrival brings other signs that spring is now in the ascendancy, as more and more flowers bloom and early butterflies emerge from hibernation. A run of a few warm days delivers up comma, brimstone and small tortoiseshell, all of which will have spent the winter in their adult form, a trait shared by just a handful of our butterfly species. For the brimstone this may have meant wintering hidden in among the dense hedgerow ivy, while for comma and small tortoiseshell the log pile or an outbuilding might have been the site chosen. There is also a growing dawn chorus, as new blackbird, dunnock, song thrush and great tit voices join those already singing. I am lucky here; even though I live in a town, my house occupies a quiet spot near the cemetery and is set back from the nearest road. This reduces the traffic noise significantly, something that is good for me and good for those birds busy establishing breeding territories. Studies across the world have highlighted the impact that traffic noise can have on birds using song to set up territory and attract a mate. Many struggle to be heard, their songs drowned out by the noise of passing traffic, greatly reducing their chances of making a successful breeding attempt.

Sensitivity to traffic noise varies with species. Traffic sounds tend to be low in pitch, their impacts on birds most pronounced in those species whose songs are of a similar pitch. If your song is loud and high pitch, it is more likely to penetrate the background noise generated by passing cars and lorries. Some birds attempt to counter the impacts of noise pollution by either changing the times at which they sing, often moving to a quieter time of day, or by altering their song. It has been suggested that the robins heard singing at night in London and other cities only do so in order to avoid the daytime noise. Blackbirds living near Madrid Airport sing more very early in the morning before the first flights leave, but also drop some of the lower-frequency components from their songs. Other effects of background noise can be more subtle, for example masking the calls of hungry chicks to the extent that their parents are less stimulated to feed them.

One of the most obvious songs of these early April days is that of the song thrush, at least two pairs of which have established territories within earshot of the garden. The thrush's song has a clarity that seems to cut through the background chorus of other birds. Although a male song thrush may have a hundred or more phrases in his repertoire, some of which mimic other sounds or the calls of other species, it is the manner in which these are repeated that brings his song to our attention. The phrases tend to be repeated in blocks of three, with some of the phrases strongly emphasised. There is an energy contained in these repeated phrases, something captured so vividly by Lord Tennyson in his wonderfully evocative poem 'The Throstle', written in the 1850s while he was living at Freshwater on the Isle of Wight. The presence of the singing song thrush is strongly felt in the poem's opening verse, which reads:

Summer is coming, summer is coming.
I know it, I know it, I know it.
Light again, leaf again, life again, love again.
Yes, my wild little Poet.

Few poems capture the song of a bird quite so well as this one, and while other poets use the thrush's song as a vehicle to mark the turning of the year, or to write of human things, only Tennyson's poem truly celebrates this wonderful songster and his ear-catching song.

One morning, midway through the month, a new sound is added to the chorus. Rich and melodic, it is the song of a male blackcap and my first of the year. The song sounds almost conversational, a fluid sequence of notes that hangs in the damp dawn air unanswered. Keen to see this new arrival I venture out, cautiously opening the back door and mindful of my footsteps as I edge down the garden and nearer to the bird responsible for this beautiful hymn to the unfolding spring. He is somewhere towards the back of the garden, in an area of bramble that has held singing blackcaps in each of the last two years. Is this the same individual, or is it instead the temporary occupation of what appears to be a suitable territory by a passing bird? The bird falls silent. Worrying that I have disturbed him, I halt my progress, immobile, ears straining to hear the soft 'two-pebbles-tapping-together' alarm call that would speak out at my intrusion. A moment passes and then he begins to sing again, this time from the holly tree that marks the boundary of our land. I can just about see him, a patch of brown half obscured by the dense holly growth. It is wonderful to have a singing blackcap present here, even if only for a short while. In both of the previous years a male was present for just a few days, reinforcing my suspicion that this is not a suitable breeding territory after all, the bramble big enough to hold a nest but the site lacking sufficient food for a breeding pair. There is a better habitat nearby, several gardens over in a bigger plot, where trailing ivy meets a tangle of bramble and a stand of last year's nettle stems. If they are to breed anywhere, then it will be there.

One bird that I suspect is breeding in the garden is the male wren whose piercing trill has been much in evidence over recent days. The volume of his song, both in terms of the amount delivered and its loudness, is remarkable. The rattling trill is pushed out energetically, each burst running over several seconds; there is a brief pause and then the song is repeated. It has been shown that a male wren, singing soon after dawn, will average some 130 to 200 songs in the course of an hour, a behaviour that will occupy a good twenty minutes of his time during that period. The male wren can produce such a loud song because he possesses an amazing vocal organ, the syrinx, common to birds and located at the bottom end of the trachea, or windpipe. Like all other vertebrates, birds possess a larynx, which is an important valve in the respiratory system. However, while the larynx of mammals, reptiles and some amphibians also functions as a vocal organ, in birds this role is delivered solely by the syrinx. The selective forces that lay behind the evolution of the syrinx are unclear, but it is worth noting that, as a group, birds have the longest necks of any air-breathing vertebrates, and a long neck requires a long trachea. Since the trachea is essentially a tube, placement of a vocal organ at its base can utilise the resonant character of the trachea and increase the volume of sound produced. The syrinx, then, is a highly efficient sound box, capable of delivering the incredible volume and variety of song and calls used by birds.

Song is not the only thing that a male wren uses to attract a mate. Like the blackcap singing earlier this morning, the wren is one of several garden birds in which the male builds a number of 'cock nests' with which to woo a prospective partner. Male wrens with more cock nests in their territory are more likely to attract a mate. Some males may build up to a dozen nests in their territory, which takes a considerable amount of effort. As a male wren gets older so the number of nests that he builds each season also increases. However, putting age to one side, some males are simply more accomplished nest builders than others, and reap the benefits of this through the number of females they attract.

In both wren and blackcap, the cock nest is rarely a fully finished affair; instead, it is probably just finished enough to demonstrate the male's abilities. Many of the blackcap nests I find early in the season are cock nests, made from a few dozen grass stems woven into a simple basket. Like a basket, the blackcap nest has 'handles' that attach the structure to the surrounding vegetation. The presence of these handles can help with identification of the nest, since they are absent in the rather similar-looking garden warbler nests often found in similar sites. Some of these blackcap nests are adorned with the fluffy white of plant down or silk stolen from spider webs, but the nest cup is left unlined; if used, then the eggs will nestle in among a lining of fine grass stems. The male wren's cock nest is also left unlined, but a feather lining will be added to the chosen site once the two birds are paired. In wrens, lining the nest is the responsibility of the female alone. One of the other advantages that a male wren can gain from having multiple nests is the ability to attract more than one female, an approach that can see a successful individual polygamously paired with two, three or even four partners. Since the male rarely visits his mate while she is incubating, this appears to provide additional scope for his infidelities. He does, however, take a role in provisioning the chicks and guarding them when they first leave the nest. I know of one of our resident male's nests, which is built into the ivy on a wall at the back of the garden. It wasn't lined when I found it, so maybe it will not be used this season, but I will check again in a couple of weeks just in case.

Wren nests are constructed from leaves, moss, dry grass and other plant material, all of it sourced from nearby. The males can be pretty obvious when building, the small bird with a vast ball of material in its beak. The nest is dome-shaped, about the size of a melon, and with a rigid entrance hole that is formed from tightly woven material. Usually they are well camouflaged, blending into the surrounding vegetation, but occasionally they may be built from material very different in colour or texture to that in which the nest has been placed. One nest that I found a few years ago was made mostly from dead bracken, which would have been fine had the wren placed the nest in the abundance of bracken that covered the site; instead he chose to build his nest in the one holly bush present on this particular patch of rough ground. A pale rusty-brown bundle of vegetation set against shiny green foliage was hardly subtle, and unsurprisingly the nest was never used. A domed nest offers additional protection to a nesting bird from some nest predators, though not all, and it also makes it less likely that the nest will be parasitised by a cuckoo. Having said that, I did once find a cuckoo egg in an abandoned wren's nest at a heathland site popular with both cuckoos and wrens.

Song and the number of nests built are not the only signals of a male's quality; plumage characteristics can also be a sign of quality and, more often than not, social status. In many garden birds, the male plumage is more brightly adorned than that of the female. A brightly coloured female siting on a nest may be more likely to be spotted by a predator, so there is likely to be a selection pressure on female plumage towards colours and patterns that blend in with the surrounding vegetation. Males, on the other hand, can afford to be more brightly coloured. Watch your garden birds and you will see there are some species in which the males and females look very different, think of chaffinch and bullfinch for example, while in others the differences are more subtle or, to our eyes, absent altogether. The dunnock is one such species, the two sexes identical in appearance, and separable in the hand only when breeding: the female with her brood patch (used for incubation) and the male with his swollen cloacal protuberance (his reproductive organ). In such cases, where the two sexes are similar in their appearance, it is only by watching the behaviour of the birds that you can gain an insight into their likely sex. Song is often a good indicator of sex but it is worth noting that in some species, such as robin and wren, the females may also sing at certain times of the year. Behaviour is another useful indicator.

The breeding season has clearly begun for the local dunnocks, the birds have been singing for some time, and I often see three individuals interacting with one another in a manner that suggests they are part of a breeding pair, or to be more accurate a ménage à trois. Their presence hints at the wide range of mating systems to which this rather unassuming bird subscribes. This can include monogamy (one male and one female), polygyny (one male and several females), polyandry (one female and two or more males) and even polygynandry (several males with several females). Male and female dunnocks have their own, largely independent, territorial structure. Because the male territories are larger than those of the females, this tends to mean that each male territory is likely to overlap with more than one female and her territory. In addition, and this is where the complexity comes in, some male territories are shared by two males, one of whom is dominant over the other. The two males work together to defend their territory against other males, but they also compete with each other, as is evident by the fact that the dominant male seems to spend a disproportionate amount of his time guarding his female(s) from the other male's attentions. For the female, having two males to help rear her chicks can be advantageous, so it pays her to court the attention of both.

Alongside the complexity of these arrangements is an equally complex series of behaviours, largely centred around courtship and parentage of the resulting chicks. Prior to copulation, a female dunnock will crouch low in front of her mate, soliciting his advances through a display that involves fluffing up her body feathers and lifting her tail to expose her reproductive organ, known as the cloaca. The male then responds to this by approaching the female from behind; hopping from side to side, he then begins to peck at her cloaca. This pecking stimulates the female to eject some of the sperm from previous matings. Shortly after this, the male mates with her. With some of the sperm from earlier matings ejected, this behaviour increases the male's chances that he will be the father of her offspring. This is important because it is likely that the female dunnock will have sought similar copulations with other males, either those in neighbouring territories or the subordinate male in this territory. Matings solicited with individuals other than your mate are known by scientists as 'extra-pair copulations' (copulations outside of the pair bond); these can increase the genetic diversity of the female's clutch of eggs (good for the female) or share a male's genes across several clutches (good for the male), both of which can be advantageous. Interestingly, a subordinate male dunnock will only help to provision a female's chicks if he has had the opportunity to mate with her, again emphasising some of the behavioural advantages that come from cheating.

While the dunnocks might have found an approach that works for them, cheating on one another, they have yet to find a response to a rather different piece of cheating, namely when a cuckoo lays its egg in a dunnock nest. Dunnock is a relatively recent host for cuckoo and appears unable to recognise a cuckoo's egg when one is present in its nest, despite the very different appearance in their eggs. A number of other garden bird species may occasionally host a cuckoo chick in their nest, including pied wagtail, robin, blackbird, spotted flycatcher and greenfinch. I have seen cuckoo eggs in both dunnock and robin nests but, as yet, not in these other species.

* * *

By mid-month there are plenty of early season blooms in the garden, many of which are being visited by the first flush of insects. Early season nectar is important for bees, newly emerged from hibernation, and among these it is the hairy-footed flower bees and bumblebees that tend to catch my attention. The bumblebee queens split their time between the blooms that will fuel them and the search for a suitable nest site, the latter characterised by a low-level flight that sees them follow the contours of the ground in search of a likely burrow. Not all of the bumblebees nest at ground level: the tree bumblebees – characterised by their dark bodies, ginger thorax and white-tipped tail – often occupy the hole-fronted nest boxes favoured by garden-nesting blue tits. This bumblebee species is a relatively recent colonist, having arrived in Britain in 2001 following a well-documented range expansion on the Continent.

A warm April morning will often find me out in the garden, net in hand, watching the flowers for visiting insects. Their presence is another reassuring sign that winter is behind us and spring firmly established. In addition to these larger, more obvious insects, there will be many others emerging to continue the cycle of life that is played out through successive generations. As the days lengthen, so this invertebrate bounty will feed the coming generation of garden birds. The long days of the northern summer help to drive this abundance of invertebrate life and, in turn, it is this seasonal abundance that supports the millions of migrant birds that are now arriving from wintering grounds located far to the south in Africa. Species such as house martin, swallow, willow warbler, whitethroat and spotted fly-catcher, all of which most likely evolved within Africa, now fly many thousands of miles to take advantage of the insect populations of our more northern summer.

The first of these migrants will have arrived back in March; a vanguard of chiffchaffs, blackcaps and wheatears, but it is in April that the numbers really begin to pick up. The first half of the month sees the arrival of house martins, willow warblers and redstarts, but it is not until several weeks later that we see our first whitethroats and sedge warblers. Finally, as April shifts into May, the first cuckoos, swifts, turtle doves and spotted flycatchers appear. The timing of these spring arrivals is captured in the notebooks of birdwatchers and the logs kept by a scatter of bird observatories located at sites around Britain's coastline. Comparison of records collected in the 1960s with those collected more recently reveals a shift in the pattern of these spring arrivals. Most of our familiar summer migrants are now arriving in the UK a week to ten days earlier than was the case in the 1960s, clear evidence of the impact of a changing climate. Not only this but some are also leaving our shores later, their breeding season effectively extended because of climate change. The species that have advanced the timing of their arrival tend also to be those whose breeding populations have shown a more positive pattern of change over the same period. Not all of our summer visitors have fared so well, however, and several – including cuckoo, turtle dove, swift and spotted flycatcher – are among our most rapidly declining bird species. One thing is certain: there are fewer insects around now than was the case just a few decades ago.

The widespread decline in our invertebrate populations has only been brought to national attention relatively recently, raising concerns over the knock-on damage being caused to the other species that feed on them and to the economic fortunes of agriculture and horticulture, dependent on pollinators for crop production. While the loss of invertebrate populations is linked to changing habitat management and the use of pesticides more widely across the landscapes, gardeners are not without blame. Many gardeners still use pesticides in their gardens, or favour plants that offer little in the way of resources to visiting insects. Fortunately, there is growing recognition of the value of adopting more wildlife-friendly approaches to gardening, and this has to be good for garden invertebrates and the other species that are dependent upon them. Such approaches are centred on an understanding of the links between species, the dependencies that hold our ecosystems together. Treat one part of the system badly and the damage may be felt across the system as a whole.

* * *

The apple trees come into flower as the month nears its end and I keep my fingers crossed that we don't suffer from a late frost, which would damage our chances of fruit later in the year. The presence of the blossom reminds me of my parents' garden and, as a child, watching the arrival of the bullfinches. These were regular visitors to the garden, arriving in spring to feed on the cherry and apple buds, returning in summer to tweak dandelion seeds from their plants, and present in winter at the bird table. The preference for buds in late winter and early spring, each bud easily dealt with by the bullfinch's robust little beak, once brought them into conflict with the horticultural industry. These beautiful little finches have been regarded as a serious pest in commercial fruit orchards because they can consume buds at a rate of thirty per minute and because they favour the flower buds over leaf buds, damaging the crop yet to be formed. Flower buds are more nutritious than leaf buds, hence the preference shown. Research into bullfinch feeding preferences has also revealed that plum and pear trees are favoured, but that the birds may also feed on currant, gooseberry, cherry and apple. Because of the resulting damage, bullfinch has featured on a 'general licence', which enabled the owners of commercial orchards to control their numbers.

The bullfinch is a species that is easy to overlook. It is a shy bird, with a quiet call and an unassuming song, that favours scrubby woodland habitats and dense nesting cover. Bullfinches tend to only visit those rural and suburban gardens that are connected to nearby woodland and scrub by hedgerows or shelterbelts. I have yet to see them visit the feeders here, but I have heard them nearby on occasion. Nationally, bullfinches appear to be making greater use of hanging feeders now than was the case twenty years ago, which ties in with a wider increase in their breeding populations. This increase, which began back in 2000, represents a change in fortunes; numbers had previously been in decline since the mid-1970s and the population is now a third smaller than it was before the decline kicked in, so any sign of a recovery is to be welcomed. Other factors contributing to the increased use of garden feeders may have been the introduction of new seed mixes, which are easier to handle, and the development of circular perches, which provide a different perching position from the traditional straight perches that used to be a feature of garden bird feeders.

Despite the fact that the bullfinch has such a soft and undistinguished call, it is widely known to be a particularly skilful mimic. It is for this reason that this finch was once a firm favourite with those interested in ornamental caged birds. Not only did a male bullfinch look the part, he also had the ability to learn and mimic fashionable tunes, a trait that could then be shown off to household guests and other visitors. From at least the 1500s through to the Victorian age, bullfinches were caught and imported to Britain for this purpose. Many of the birds were trained by their owners, who whistled a tune or used a special 'bird-flute' in order to train them. This practice was so widespread that it even gets a mention in Thomas Hardy's *Tess of the d'Urbervilles*.

In contrast to the bullfinch, with its secretive habits and long history of decline, the collared dove is now a particularly obvious and vocal addition to my garden bird community. Throughout much of the year, the local soundscape is punctuated by the dove's 'u-ni-ted, u-ni-ted' calls which, like those of the woodpigeon, lack any real charm. To have one calling in that monotonous tone from the roof above your bed can quickly prove tiresome. The story of the collared dove, however, is a remarkable one, the species first breeding in the UK in the mid-1950s. Those first breeding attempts took place here in north Norfolk, just along the coast from me in West Runton. They were followed by others, the birds appearing and then breeding in the scatter of towns and villages running along this stretch of coastline. By the middle of the 1960s they were present in over eighty Norfolk towns and now confirmed breeding in other English counties. The pattern of colonisation continued, with human habitation and farm premises instrumental in supporting the birds and acting as sites from which the doves could spread into other habitats. Today, the UK population is thought to number in excess of 800,000 pairs, reflecting the successful colonisation event that took place in little over half a century. The collared dove clearly found the UK to its liking and was able to exploit a vacant niche which it has very much made its own.

The colonisation of the UK was part of an even more remarkable colonisation event that had unfolded over the previous half century. At the end of the nineteenth century, the collared dove was restricted to the Balkans and a small area of Turkey. Over the following thirty years, an expansion in its breeding range took the species to Hungary, then reaching Austria in 1943, Germany in 1945 and the Netherlands in 1949. Quite what had triggered and then driven this sudden and dramatic shift in breeding range is unclear, but it is a pattern that has since been repeated in North America. Up to fifty collared doves, taken to the Bahamas by a pet breeder in 1973, escaped from their aviary. The birds bred in the wild the following year and during the 1980s they arrived in Florida, from where they have proceeded to colonise much of North America. Both in North America and here in the UK, the pattern of colonisation has involved young birds making long-distance movements to colonise new areas, followed by a process of backfilling of the gap between the old breeding range and the new front created by these pioneer birds. In both the UK and North America, garden feeding stations have been instrumental in supporting this expansion.

Another factor in the collared dove's success is likely to have been its breeding behaviour. Pairs can breed in any month of the year – they are usually the first bird that I find nesting each year – and individual pairs may make up to five breeding attempts per year. The main breeding season runs from February to October, so the adult male that has been displaying just outside the bedroom window for the last few days may already be on his second breeding attempt. Some mornings his antics prove particularly entertaining as he hops along the ridge tiles towards his mate, who walks away at a steady pace. The male, his hunched back indicative of matrimonial intent, closes in on the female and looks as if he is just about to mount her. At that moment, the female's timing impeccably judged, she takes to the wing and flies to the roof of a neighbouring house, where the whole process begins again. These birds seem to spend a lot of time on the roof, a behaviour that has earned them the name 'television dove' in Germany, because they commonly call from television aerials. Male collared doves advertise and maintain small breeding territories, using both their monotonous call and a display flight to advertise ownership. The display flight is not dissimilar to that of the woodpigeon, being a series of steep climbs, followed by a glide. In the collared dove this glide typically takes the form of a spiral, whereas in woodpigeon it is straight like an old-fashioned rollercoaster.

Collared dove nests are pretty poor affairs, a flimsy platform of small twigs placed on a ledge or well out on the branch of a tree or shrub. They seem to have a liking for the brackets used for satellite dishes and security lights, placing twigs across the frame to form a nest. Some of the nests are so poorly constructed that you can see the eggs by looking up through the bottom of the nest; needless to say, many eggs fall through the nest and smash on the ground beneath. Like other doves and pigeons, the regulation clutch size is just two eggs, which hatch after roughly two weeks. The resulting chicks remain in the nest for another two to three weeks, the whole process taking about two months from beginning to end. Even if a collared dove chick makes it through to fledging, it still faces something of a challenge if it is to go on and produce young of its own; roughly two-thirds of young collared doves die before they reach the end of their first year of life. Predators, disease and competition for food all take their toll. Given this, and the poor parenting skills of these birds, it is perhaps surprising that they have managed to colonise the UK at all, let alone so quickly. In the case of the collared dove, it seems as if this success comes from the bird's stubborn persistence.

It is only when the collared doves come down to the bird table located close to the house that I get a chance to see the more subtle colours in their otherwise seemingly rather drab plumage. While the feathers of the back and wing coverts are a mix of pale grey and soft brown, overlaid onto the dark grey flight feathers, the breast is subtly infused with pink. On the back of the neck is a narrow black collar, thinly edged with white, initially absent in young birds. But it is the eye that gets me;

this is a deep red, so deep that it almost appears black and you have to look closely to pick it out. It appears that the eye colour develops with age, young birds having pale yellow/brown eyes, and much paler pink legs than those of the adults. Over time, the eye colour deepens to a rich Rioja-red.

* * *

While collared dove society is centred on individual pairs, perhaps joining small flocks in autumn to utilise seasonal feeding opportunities, that of the local house sparrows is very much centred on the colony. Each adult male house sparrow defends his nesting cavity and the immediate area around this, but away from the cavity is the shared colony space. House sparrow society is choreographed by a hierarchy based on the social status and dominance of individual birds. The status of each male is demonstrated by the black bib that he carries on his throat, its black feathering extending down onto the top of the chest. The size of this area of black feathering reflects the male's position in the dominance hierarchy. The larger the black bib the higher up the social hierarchy the male is typically placed. While the bib appears to provide some measure of social status, what is it about these males, with their large bibs, that makes them so special? Detailed studies of active house sparrow colonies have found that large-bibbed males do proportionally more work when it comes to the division of labour between the sexes in a breeding pair. Male house sparrows help to build the nest, incubate the eggs and provision the chicks, and large-bibbed individuals do more of this than those with smaller bibs. Not only this but they also perform more of those behaviours that would be considered the riskiest, such as nest defence. Given this pattern of behaviour, it is easy to see why a female house sparrow is most likely to select a mate with a large bib and, correspondingly, why these males are the most valued and secure the highest social status within the colony. Male house sparrows also possess a second badge of status in the form of their bill colour. During the winter months, a male house sparrow's bill is the colour of horn, but come the breeding season this darkens in tone, tending towards dark brown or black. Males with the darkest bills have been found to have the highest levels of testosterone.

Our local house sparrow flock is drawn from birds that breed under the roof tiles and in the barge boards of adjacent houses. The sparrows make use of the various feeding stations scattered across the local gardens but come together in the thick ivy-covered hedge that runs up the side of our property. The sparrows can be heard chattering away, tucked deep into the cover that the hedge provides, delivering a kind of social singing that is almost conversational in nature. They also make use of the dry soil underneath two small bushes by the pond, which appears to be of a perfect consistency for dust-bathing. This is a behaviour in which the sparrows indulge on a daily basis, each bird forming a small depression as it pushes into the soil to force small particles into its plumage. Presumably the soil acts in a similar way to water, flushing feather parasites from the feathers. Once an individual finishes its 'bathing' it flies up into one of the bushes to finish preening, before joining the rest of the flock in the hedge.

Although the UK house sparrow population is now just a fraction of its former size, it is still common enough to be a familiar bird across our towns and cities. House sparrow densities peak within suburban landscapes, being greater here than in either rural or more urbanised landscapes, and this suggests that the combination of nesting and feeding opportunities available in the suburbs is more favourable. Having said this, we know from work carried out by the RSPB in Leicester, that the breeding performance of house sparrows in urbanised landscapes is restricted by a lack of invertebrate prey. Not only does this reduce the number of chicks that house sparrows fledge – especially from late-season nesting attempts – but it also results in lower survival rates of those chicks that do manage to fledge from the nest. Faced with less food, those chicks that do fledge tend to leave the nest at a lower weight. The relationship between food availability and weight at fledging appears to be behind the lower survival rates of urban house sparrow chicks. This suggests that encouraging insect populations within our gardens and other areas of urban green space, by adopting more wildlife-friendly approaches, could improve the fortunes of the house sparrows that breed alongside us.

The chances are that, this late in the month, our local house sparrows will already be incubating a completed clutch of four or five eggs, laid in a relatively untidy nest located in one of the many cavities these old houses offer up. I suspect that a few pairs might even nest in the thickest part of the old hedge, away from where the colony gathers to chatter communally, but so thick is the ivy cover that there is little chance of discovering where. House sparrow eggs hatch within two weeks of incubation beginning, so it will not be very long now until the birds will have hungry mouths to feed. Here's hoping that April's end will offer up some fine weather and a wealth of invertebrate food for these engaging little birds.

May

There is a moment, after days of rising expectation, when I see my first swift above the garden. It is not my first of the year, but it is the bird's presence here that marks their arrival for me. It is the few dozen pairs locally, occupying those older properties with access under their roof tiles, that I view as my birds, and my hunger for their return is at last satisfied. These are the swifts that I will see hawking above the garden over the coming weeks and whose drawn-out screeching calls will draw my gaze upwards with undiminishing enthusiasm. There is something so energising about their presence, these birds that live on the wing and spend only a few short weeks with us. The knowledge that they have flown from Africa, spending the first part of the winter in Central Africa before pushing southeast to Mozambique for the remainder, makes their presence here in the spring all the more remarkable.

It is only recently that we have uncovered the mysteries of swift migration, joining up the dots that came from occasional recoveries of ringed individuals far away from their breeding sites. Now we have the tracks recorded by tiny tracking devices that link together the breeding sites here in England with the great arc of movement that sees the swifts cross the African continent from centre to southeast, before looping back into West Africa in late winter ahead of a return to the northern summer. These are truly remarkable birds, playing out life on the wing and only landing to raise their young. Swifts do not make their first breeding attempt until they are at least two years old, often older, which means that once they fledge from their natal site they may not touch down again for another four years. They feed, mate and even sleep on the wing. To see them in flight underlines that they are masters of the sky; to see them in the hand – when monitoring them at their nest sites – underlines how unsuited they would be to the way of life typical of most other birds. The long wings and short legs would leave a swift nearly helpless were it to be grounded.

Over the next few days, broken only by a short spell of poor weather, I see the swifts every day and go from seeing just that first solitary bird to sixteen birds in the air together. It finally feels as if summer is here and, with a nod to the poet Ted Hughes, that all is right with the world and the globe still turns. A neighbour has two swift nest boxes on his property and I would love to see the swifts use them. It might be a bit of a wait, however, as the adoption of new sites appears to be a slow process. Swifts are monogamous, their pair bond maintained from one year to the next even though the members of the pair probably do not associate with one another during migration or on the wintering grounds. The birds show strong fidelity to their breeding colony and return to both the same colony and the same nest site in successive years. Young birds tend to settle at a colony different to the one in which they were reared, thereby reducing the chances of inbreeding.

When the birds first arrive we tend to see them feeding over the local fen, but as the days warm they spend more and more time above the town. In with the adult swifts are many younger birds, not yet of breeding age, boosting the numbers present in the air on these warm evenings. Those adult birds that have nest sites in town will use these for roosting overnight, the pair of birds crouched together in the darkness of the nest cavity. Younger birds roost on the wing, gathering in early evening in parties that bunch ever more closely together, as they ascend into the sky. Although the paired adults may join in with this evening ascent, they soon break away to return to their nest cavities, the responsibilities of breeding a temporary shackle to the earth below. The young birds and non-breeders continue to rise, disappearing from view into the darkening evening sky. Up and up they go, gaining as much as 2,500 metres in altitude and taking them well beyond our gaze.

Until relatively recently, it was thought the birds making these evening ascents, termed 'vesper flights', stayed at these incredible altitudes through until dawn, sleeping on the wing in a layer of warm air. More recent studies paint a somewhat different picture. The use of Doppler weather radar, capable of detecting individual birds and the pattern of their wingbeats, has revealed that swifts make two ascents during the course of the night, one in the hour after sunset and one in the hour before dawn. That the birds make two ascents, each a mirror image of the other in relation to dawn and dusk, suggests that these flights are not simply to roost but instead serve some other purpose. While the evening ascent and morning descent are made as a group, the evening descent and morning ascent are carried out alone, each bird making its own way back down to earth and back up again. Researchers now suspect that vesper flights are used by the birds to orientate themselves and to identify the weather patterns, such as frontal systems, coming their way. Swifts range widely, and it has long been known that they will forage over vast distances in order to avoid the poor weather conditions linked with particular weather systems. In some way, it seems, the swifts are using these flights to forecast the weather, the knowledge gained shaping their behaviour over the following days.

One feature of the swift flocks that draws attention to their presence is the periodic screaming displays in which groups of birds barrel across the sky, dashing past at rooftop level, while delivering the characteristic drawn-out scream. Many of the nest site owners will echo the call as other individuals swoop past their nest entrance, sometimes emerging to join in the chase. The groups of birds appear to make several circuits of the colony and it is thought that the behaviour has a social function, perhaps linked to colony cohesion. There have been evenings when I have been sitting quietly in the garden, the soft sounds of evening suddenly broken by the arrival of a screaming party of swifts low overhead. There is so much energy about these groups that you can't help but smile and wish you had the wings to join them.

While the swifts are busy being obvious, one of our other summer migrants is much more likely to go unnoticed. This is the spotted flycatcher, an unassuming bird, plain in colour and soft in voice. I always associate the spotted flycatcher with larger rural gardens, having failed to attract one to any of my gardens, but the species also breeds in orchards, churchyards and within broadleaf woodland. Soon after they arrive on their breeding sites these delightful little birds can be seen feeding from a favoured branch, sallying forth to grab a flying insect (often with an audible snap of the beak) before returning to their perch. The spotted flycatcher's gape is extended by a series of bristle-like feathers along each side of the bill. These are known as rictal bristles and reduce the chance that a fly or other insect caught in the bill will escape. As the name suggests, the spotted flycatcher eats a lot of flies, but it also takes a much broader range of insects, from dragonflies through to butterflies and moths. The bird isn't deterred by the warning colouration of wasps or bees, dealing with the sting by rubbing the captured insect against a branch until the sting is removed. It is clear that the spotted flycatcher recognises the challenge that stinging insects present, since it will only feed on them if other prey species are less readily available.

We still have several pairs locally, all within walking distance of the house, but this is a bird that is easily overlooked because of its soft call. I suspect that many of those fortunate to have a spotted flycatcher nesting within their garden will be unaware of the bird's presence. The spotted flycatcher's nest is almost invariably placed on a small ledge, either provided by the brickwork of a building or by vegetation, such as a climber like wisteria. Unlike the robin, which likes its nesting cavity to be well hidden, the spotted flycatcher likes to have a good view from its nest, which is why the open-fronted nest box designs used for this species have such a low front section. Breeding spotted flycatchers are unobtrusive in their behaviour but individuals may be seen hunting for flies and other insects from a favoured perch. Successful breeding is dependent upon the birds finding sufficient prey for their brood of four or five chicks, and larger insects appear to be of particular importance. There is a suggestion that adult flycatchers are more likely to eat small insects themselves, only taking larger prey back to the nest, so a poor summer for large prey species may make things particularly difficult for spotted flycatcher chicks. Interestingly, all of the remaining local pairs known to me breed along the river, a habitat that is likely to support better insect populations than is the case here, where I am surrounded by other gardens.

Spotted flycatcher pairs often use the same nest site, or one nearby, in subsequent years, so if you have them breeding there is a good chance that they will return the following year. The return is, however, dependent upon a successful migration to wintering grounds in Africa. It is only very recently that the location of these wintering grounds has been revealed. Evidence from the recoveries of ringed birds, coupled with observations from Africa, suggested that spotted flycatchers wintered south of the Sahara and probably in West Africa, but the use of tiny tracking devices has revealed that the wintering grounds of those spotted flycatchers breeding in the east of England are actually located in central Africa, in Angola. Knowledge of where the wintering grounds are located, coupled with information on the migration routes and stopover sites used by these birds, can help to identify factors that might be implicated in the changing fortunes of this Red-listed species, whose UK breeding population has declined by 89 per cent in just fifty years.

The reasons for the observed decline in spotted flycatcher populations are unclear. There is some evidence for changing levels of nest predation here in the UK, with rates of nest survival actually higher in gardens than they are in woodland or farmland habitats. However, changes in the availability of insect prey may also be having an impact, and other factors may be impacting the birds during migration or on their wintering grounds. Spotted flycatcher is one of a suite of migrant species, all wintering within Africa's Humid Zone, that show long-term declines in their UK breeding populations. This suggests that the chances of a spotted flycatcher using the nest box that I have erected for it remain slim. However, I feel that I am doing my bit, even if the empty nest box hints at the continued decline in this once much more common species.

* * *

In the second week of May, a friend calls to tell me that one of their nest boxes has been plastered with mud, the lid now glued to the front of the box by a line of mud placed inside, and the entrance hole narrowed in a similar fashion. This, as they recognise, is the work of a pair of nuthatches and I tell them that they are fortunate; both to have nuthatches nesting in their garden and because the nuthatches have used mud rather than dog faeces to seal the box, which they can sometimes do. They seem less than amused by this second comment. The use of mud to narrow the size of the entrance hole is a sensible strategy, preventing the entry of larger birds that might oust the nuthatch pair from its chosen site. The use of mud elsewhere is probably a response to the slither of daylight that might suggest a weak point in the cavity's exterior. Plastering the gap with mud fixes this but it also makes it difficult to access the box for nest monitoring, which is a pity given the relatively few BTO Nest Record cards submitted for this species annually.

Dog faeces aside, nuthatch nests can be rather attractive affairs. Lacking a woven nest cup of the sort more typical of small birds, the eggs are often laid onto a loose lining of flaked bark, taken from pine, larch or birch. This gives the nest a rather picturesque quality, the warm brown tones of the bark flakes complementing the pale speckled eggs.

Nuthatches are strongly territorial birds, often defending their breeding territory throughout the year, and quick to eject newly fledged young from it once they have left the nest. They can also be aggressive at the feeding station, the large chisel-like beak a substantial deterrent to other species. With the exception of the great spotted woodpecker, the nuthatch is dominant over just about every other species that might attempt to share a feeder or bird table with it. Some indication of just how antisocial these birds are can be seen from the text introducing the section on 'Social Behaviour' that appears in the multi-volume work seen as the definitive account of the birds of our region, known to birdwatchers as *Birds of the Western Palearctic*. The nuthatch account begins by saying the '[m]ost notable aspect of [their] social behaviour is [the] markedly unsociable behaviour', before going on to note how even newly fledged individuals readily threaten their siblings if they come too close.

Although we do not have breeding nuthatches in our garden, we do see them occasionally in the winter months, typically visiting the bird table or one of the hanging feeders. It is when you see them away from the garden that you get a sense of how they would more naturally forage for food. With strong feet and claws, nuthatches are able to forage on tree trunks for insects and spiders. During the autumn, when tree seeds become available, they may be seen to carry an acorn or hazelnut to a suitable crevice, forcing it into place before striking it open with the robust beak. This behaviour is behind one of the old local names for this bird of 'nut jobber', 'job' being an old English word meaning 'to stab with a sharp instrument'. The bill may also sometimes be used to chisel at loose bark, the intention being to gain access to any insects or spiders that happen to be hiding underneath.

Almost as good as having a nuthatch in your nest box, or so I tell myself a couple of days later, is finding a chaffinch nest in the ivy that has grown up over the old hedge line along the edge of the garden. The nest is beautiful, with a deep cup and a bulbous outline. It is constructed from moss, into which has been woven rootlets and grass. It is soft green in colour but has been adorned with patches of grey lichen, wool and plant down. Within the cup, which is lined with feathers, are three eggs, cold to the touch, so I suspect that the female is still laying her clutch. I make a note of the nest and its contents in my notebook, hopeful that these will form the beginnings of a record for the BTO's Nest Record Scheme. Chaffinches, goldfinches and greenfinches may all use gardens for nesting, and over the years I have found the nests of all three in my garden. While the chaffinches have tended to be found in ivy, those of goldfinches are more usually located in a tall shrub or fruit tree, the nest usually placed in the crook of a branch. They also tend to be placed higher off the ground. Greenfinches prefer evergreen cover and larger ornamental conifers seem to be a firm favourite. Unlike the two other finches, greenfinches often nest in loose colonies of four to six pairs, the birds less clearly territorial and only really defending the area immediately around the nest. Chaffinches are territorial and the male's song-posts mark the boundaries of the territory in which our chaffinch nest is located.

A week later and I am working in this bit of the garden, so I make a careful second visit to the chaffinch nest. Softly tapping the ivy with a cane as I approach, I hear the flutter of a bird, the female slipping off the nest much as she might do if accidentally disturbed by a browsing herbivore. A quick glance reveals that the clutch now numbers six eggs, warm to the touch, indicating that incubation has begun. Other visits will follow, and I plan the first of these by calculating the likely date on which the eggs should hatch. Like most small birds, chaffinches typically lay one egg a day, and they usually begin incubation once the penultimate egg has been laid. With an incubation period of just under two weeks, I calculate when my next visit should fall, just after the eggs have hatched. This will provide a measure of hatching success and reveal whether the nest has survived to the chick stage. Hopefully, I will then make two further visits, one to ring the chicks just as their flight feathers are starting to emerge, and a final visit to determine whether the young have fledged successfully. Each piece of information gained provides valuable data for the researchers seeking to explain how our bird populations are doing and what is driving any change in numbers evident in annual surveys of their populations.

While my opportunities to follow the progress of this chaffinch nest are deliberately limited by not wanting to increase the amount of disturbance around the nest, emerging technologies are now providing us with information on nesting birds on a virtually real-time basis. The use of nest box cameras, now affordable to most people, enables you to stream the action from your nest box direct to your television, computer, tablet or smartphone. Perhaps stimulated in part by BBC's *Springwatch* series, many garden birdwatchers use nest box cameras to watch the progress of nesting attempts. The real-life soap opera that unfolds as the season progresses can be something of a rollercoaster watch, as the birds face the challenges of raising a brood in a garden setting. A run of poor weather, or a poor year for favoured insect prey, and a whole brood can be lost, even at a relatively late stage. The use of these cameras provides an opportunity for some interesting citizen science, with the possibility to secure data on feeding rates, chick growth and fledging behaviour. They also provide an opportunity to link the chicks from a nest box, watched on camera from egg to fledging, with the sudden arrival of newly fledged young on the feeders outside the kitchen window, something that provides a sense of relief and satisfaction that 'your' nesting blue tits have succeeded.

* * *

By late May, use of the hanging feeders has fallen, reflecting the fact that most of the birds are now settled on eggs or have young chicks in the nest. It will be insects, spiders and other small invertebrates that are the food of choice for the garden's birds over the next few weeks, and I will not see an increase in feeder use until there are young birds newly fledged from the nest. The garden nest boxes, many of which I made myself, now host broods of hungry blue tits and great tits, whose parents will be working hard to find sufficient caterpillars with which to feed their young. I try to make sure that I have a range of different box designs, broadly split between those with a small circular entrance hole ('hole-fronted') and those that have a much larger opening ('open-fronted'). Within the hole-fronted designs it is important to vary the diameter of entrance-hole provided, restricting access to some of the boxes to just the smaller species, such as coal tit. I have found that an entrance hole diameter of 25 millimetres is ideal for blue tit and coal tit, while 28 millimetres works for great tit and 32 millimetres for house sparrow.

Of course, sometimes a blue tit will occupy a nest box with a large aperture, but it then runs the risk of being evicted by a larger, more dominant species. One consequence of this is that, very occasionally, one may encounter a mixed brood containing both great tit and blue tit chicks. I encountered such a brood a couple of years ago at a woodland site that I monitor, with two blue tit chicks in among eight great tit chicks. The adults visiting the box to feed were great tits, so it looked as if they had evicted a blue tit pair that had just started laying their own clutch of eggs. Since these small birds do not begin incubation until the clutch is just about complete, the blue tit eggs would have only started their development once the great tit had initiated her own incubation. The two species have a similar incubation period of roughly two weeks, so the eggs of both species would have hatched at about the same time. Just as great tits may take over cavities used by blue tits, so both species may take over a site being used by the smaller, and less dominant, coal tit. This may be one reason why coal tits often end up occupying cavities that are located close to ground level. Such sites probably carry a higher risk of predation and so are less favoured but, faced with competition, they may be all that remains available. An experiment has been carried out whose results support the hypothesis that it is competition rather than preference which results in the use of low-level cavities by coal tits. By placing boxes at different heights in the absence of competition from more dominant tit species, the researchers found that coal tits actually favour sites that are, on average, located some nine metres off the ground. When faced with competition, they end up using those boxes located lower down.

Once a tit has secured a nest box or natural cavity, built its nest and laid its clutch of eggs, it will begin incubation. Female tits often roost in the cavity while egg-laying is still in progress – as mentioned earlier, small birds typically lay one egg a day until the clutch is complete – and they will often position themselves over the eggs while roosting. You might think that this would lead to incubation beginning, but even though the roosting bird may be in contact with the eggs she is not, at this stage, generating enough heat to initiate the process. The heat that is needed to incubate the eggs comes from the female via her brood patch. This is an area of bare skin that develops prior to incubation and following the loss of its feather covering. As the brood patch develops so the blood supply to the area is increased and it becomes warm to the touch. Studies demonstrate that a female great tit is able to maintain the surface temperature of her clutch at around 35°C. The female will take breaks from incubation, leaving the nest for five or ten minutes, typically every thirty to forty minutes. In some species, the male may visit to cover the eggs (though not incubate) while the female is away. The female will also brood her chicks when they first hatch from the egg, because at this stage they may be unable to regulate their own body temperature.

While those small birds that nest in the open tend to have small broods of typically four or five chicks, cavity-nesting species tend to have more young. This appears to reflect the increased security that a cavity provides, enabling the chicks to remain in the nest for longer, but there are also other factors at play. Both blue tit and great tit are single-brooded, making just the one nesting attempt each year. This 'all your eggs in one basket' approach means that if conditions are good then a pair of nesting tits stands a good chance of fledging eight or more chicks. However, if conditions are poor then they are unlikely to fledge any young that year. In contrast, a blackbird pair will typically make two or three nesting attempts during the season, each with three or four young. The blackbirds will need to be successful across all of the attempts if they are to match or better a successful blue tit nest. However, if one of their attempts fails then they still have other opportunities to fledge at least some young over the course of summer. For the blue tit, there is no margin for error.

Open-nesting species face an increased risk of predation, their nests accessible to a broader range of nest predators and also more susceptible to the elements. This may be one reason why the development time of open-nesting young can be surprisingly rapid. It continues to astound me how a brood of blackcap chicks, seen naked and newly emerged from the egg at the start of a week, are sufficiently grown and feathered to leave the nest just over a week later. The pressures to develop in preparation for departure also mean that some attributes are prioritised over others. For example, much of the body feathering is absent when the chicks first leave the nest, the resources instead directed towards the development of the flight feathers, which are more important. Although capable of leaving the nest, many open-nesting chicks will continue to be dependent on their parents for a week or more after leaving the nest. That this is the case becomes quickly obvious if you have newly fledged blackbird chicks in your garden, harassing their parents for food. During the first few days after leaving the nest, blackbird chicks tend to remain in thick cover near the nest, calling for food. Over the coming days, they venture farther afield and soon make their first appearance on the lawn or near the bird table. The chicks continue to utter their call, becoming particularly vocal if a parent is nearby with a beak full of food.

While the threat of predation may be a driving force behind the rapid development of blackbird chicks, the threat is not removed when the chick has left the nest. Newly fledged blackbirds, naïve in their behaviour, may be taken by the cats and sparrowhawks that hunt the garden and those of our neighbours. In addition, chicks may be killed by flying into a window or greenhouse, or through getting caught in garden netting. The mortality rates of these young birds are high, a pattern repeated across most garden bird species, and it is easy to see why blackbirds have to make several breeding attempts each year if they are to see chicks survive through to the following breeding season. That we are not knee-deep in blackbirds or blue tits reflects the very high levels of mortality that these small birds face during their rather short lives.

It is not just through their behaviour that the young blackbirds give themselves away. Their juvenile plumage is rather different from that of their parents, a pattern seen in many other garden birds. A juvenile blackbird, newly fledged from the nest, has a warm brown appearance. This is due to the rufous-coloured markings that run through the juvenile body feathers; streaks of colour that are most obvious on the small contour feathers that run over the top of the closed wing but which can also be seen down the back of the head and neck. The background colour of the feathers is closer to that of an adult female, though look closely and you may notice that this colour is darker in some individuals than in others. Those with the darkest background colour are likely to be young males, something that will only properly reveal itself once the bird undergoes its post-juvenile moult. In blackbirds, this moult tends to happen four to six weeks after a chick has left its nest.

Watchers of garden birds sometimes comment on young birds, suggesting that they are larger than their parents. While one might imagine that these newly fledged youngsters are carrying some extra weight as an insurance against those first few days of independence, the appearance is a deceptive one. That 'larger' size comes not from extensive fat reserves but from the looser body plumage, which is initially sparser than that seen in an adult bird and more open in nature and less strongly contoured. It is impossible to see just how sparse the body feathering is unless you happen to come across a dead fledgling or regularly handle young birds as a BTO ringer. In the hand, and with a soft pulse of breath onto the bird's flank to expose the feather bases, you soon see how few feathers make up the body plumage in these young birds. However, in among these fully formed feathers are the first signs of others, just emerging from the skin and still 'in pin', each held within its sheath.

The young of most garden birds have a recognisable juvenile plumage, one that is replaced when the individual undergoes its first moult. This first moult usually sees the bird replace its body plumage: those feathers covering the head, back, flanks, chest and belly. In addition, some of the contour feathers on the wing are usually replaced, including some of the coverts – the feathers that sit immediately above the longer flight feathers. You can see the results of this post-juvenile moult quite clearly in young male blackbirds, which show the dark body plumage of an adult but still have the brown flight feathers of a younger bird. It is not until their next moult, the following year, that these flight feathers are replaced and the young bird attains its full adult plumage. This pattern of moult is useful to bird ringers, seeking to identify the age of a bird that they have caught. Careful examination of the primary coverts, the shape of the tail feathers (those of a young bird are typically more pointed), and other feather tracts can all provide very useful pointers to a bird's age, if you have knowledge of the typical pattern of moult exhibited by the species. Not all garden birds follow this pattern of juvenile moult, however. In a few species, including house sparrow and long-tailed tit, the young birds pass very quickly through to their adult plumage, replacing all of the key feathers during this post-juvenile moult. Within a few weeks of leaving the nest, they will look just like their parents.

Moult is energetically expensive, and it can also leave a bird vulnerable to predation and poorly insulated. This is why birds don't tend to moult all of their feathers at once, the moult of the flight feathers instead progressing in sequence so that the bird can still fly. The replacement of flight feathers in larger species often takes place over several years, an individual replacing perhaps six or seven feathers on each wing in one year, and a similar number the next. For species like the tawny owl, whose main flight feathers are replaced in this way, the difference in colour and pattern between the resulting generations of wing feathers can enable ringers to determine the age of the bird even after several years.

* * *

The young blackbirds, of which at least three are present in the garden currently, have moved away from their nest and the thick cover of the hedge in which it is located. The parents have been busy delivering food, each fledgling broadcasting its location with a characteristic call. It is not a call heard outside of the breeding season; each spring, when I hear it again for the first time, it takes my brain a few moments to process the sound and to remind myself that this is the call of a newly fledged blackbird, a sign of new avian life in the garden, and not some exotic visitor. For the first few days, the blackbird chicks remain in cover, their presence only revealed through their calls, but after a while they emerge and can be seen out in the open, perhaps on the lawn, sitting on the fence, or in attendance at the bird table while a parent feeds itself. The two adults appear to divide the responsibilities for the brood between themselves, though the female seems to spend more of her time being alert to potential danger, while the male concentrates on provisioning the chicks. Should one of the neighbourhood cats decide to sit for a while on one of the garden chairs, then it is subject to a seemingly endless series of alarm calls. Unfortunately for the blackbirds, the cat is unmoved by the attention being directed towards it by this anxious mother, leaving the rest of us to suffer the jarring torrent of abuse being delivered.

A few days later I find a dead blackbird chick near the edge of the vegetable bed, not far from the greenhouse. There is no obvious cause of death and I wonder if it has flown into one of the glass panes. Sadly, such accidents are not uncommon, for gardens hold a range of hazards for naïve and newly independent birds. Uncovered water butts and loose netting may also pose a risk to birds, and indeed to other garden wildlife. The chick is definitely one of ours, from the nest in the hedge and that brood of four, because it carries the numbered ring that I fitted a few days before it left the nest. This 'ringing recovery' will be added to the national database and provide valuable information contributing to our understanding of blackbird survival rates and how these vary between habitats and over time. More poignantly, however, the presence of the ring marks this unfortunate victim of accident out as an individual, one I have known personally. It is a life lost, not of a blackbird but of this individual blackbird, something that weighs on my mind over the coming days, as each young blackbird call heard reminds me of its loss.

Soon the remaining youngsters are joined by others of their kind; unringed, these have probably come from nests in the neighbouring gardens. Just as the expansion of foraging ranges has brought these youngsters to our garden, so our garden's fledglings have expanded theirs and we see less of them as each week passes. I hope that the three remaining chicks from our first nest of the year are now fully independent and faring well somewhere locally, out of sight but not out of mind. The first dunnock chicks of the year also put in an appearance in the final days of the month. These youngsters are less obvious and far less demonstrative than the blackbirds, but their presence at the bird table is revealed through a combination of behaviour and subtleties of plumage. While an adult dunnock has soft grey plumage around the head and chest, that of a young bird is streaked and will remain so until it moults its body feathers. Another feature useful for separating young dunnocks from adults is iris colour. Young birds have eyes that are dull olive-brown, while adults show a reddish-brown iris. This difference is only really evident during the summer months because eye colour changes quickly during a bird's first autumn. A few young dunnocks retain the dull iris colour through to February the following year, after which all individuals should show the adult colouration.

Things in the garden pick up speed from late May, as the fine weather sees a flush of new growth, growing numbers of insects on the wing and increasing numbers of young birds fledging from their nests. There is a sense that these changes are happening on an almost daily basis, and it is clear that we have left behind those early spring days, when each new change felt more significant because of the anticipation invested in its arrival. Now, it feels as if we are accelerating through spring and on towards the slow, heavy days of summer, even though in reality these are still some weeks away. I always want to have more of May, to extend this period of fresh and vibrant growth, so that I can better savour the sheer volume of activity and new life that is taking place, both here in the garden and more widely across the countryside.

June

By the end of the first week of June, the first of the year's young goldfinches has already put in an appearance at the bird feeders. Easily distinguished from the adults alongside which it is feeding, the young finch lacks the striking black, white and red facial markings of an older bird. It will not attain these until later in the summer, when it moults and replaces its head and body plumage, together with some of the feathers on its wing. Instinctively, the young bird begs for food from the adult occupying the nearest perch, fluttering its wings and calling, bill open wide. The adult, which may or may not be this youngster's parent, ignores the calls and continues to feed. The arrival of one of the local blackbirds, which has learned how to take seed from this particular feeder, scatters the goldfinches, which fly up into the clematis that covers some nearby trellis. The goldfinches are regular visitors to the feeders throughout the year and I look forward to their early summer visits, hopeful of seeing evidence that they have bred successfully.

The presence of other young birds is equally obvious over the following weeks. Vocal, though not as loud as the demanding blackbird chicks encountered last month, are the young blue tits and great tits. Again, these can be distinguished from the adults by their begging behaviour. Many a chick will perch close to a feeding adult, repeatedly calling for food while fluttering its wings. The young tits have yet to attain their adult plumage and still show soft yellow plumage tones to areas of feathering that are white or buff in the adult birds. Chick-begging behaviour triggers a feeding response in their parents, something that is most evident when the chicks are still in the nest. Here, the begging calls act alongside other features to stimulate the parent into providing food. Many chicks have brightly coloured mouths, often bright yellow, orange or pink. The chicks also have an obvious flange edging the gape, which broadens the open mouth and is usually of a paler colouration than the mouth itself. Some species, such as dunnock, the larks and the reed warblers, also have tongue spots, present for at least the first few days of the chick's life, which are also thought to encourage the adult to direct food into the waiting mouth. The gape flange can often be seen in newly fledged chicks, still present as a thickened margin to the base of the bill. Its presence seems particularly obvious, to me at least, in young house sparrows. Research has revealed that in some species, the colouration of the gape and its flange provide an indication of the health of the chick. It may be that adults respond to this by preferentially feeding those chicks most in need.

Sometimes an adult bird, taking food from a bird table or hanging feeder, will respond to the begging behaviour of a nearby fledgling, even though it is not one of its own offspring. This underlines just how strong the urge to feed a begging chick can be, perhaps even prompting an individual to offer food to one that is not of its own species. It is easy to understand this happening at a busy bird table, the adult absent-mindedly directing food to a nearby chick while busy feeding itself. However, in extreme cases an adult bird may actually deliver food to the chicks of another species while they are still in the nest. This may occur because the two nests are located close to one another, and the bird has simply heard begging calls and entered the wrong nest. Sometimes, however, where an individual's nesting attempt has recently failed, the urge to provide remains and becomes misdirected. In such cases, the surrogate continues to make feeding visits to the nest, often prompting an aggressive response from the parent birds whose nest it is.

The absolute extreme of such behaviour is that of the cuckoo chick, fed throughout its time in the nest by its adoptive parents. The level of deceit deployed is astounding, extending from the mimicry seen in the colour and pattern of the cuckoo's egg through to the behaviour of the chick itself. Rather than simply mimicking the begging calls of a host chick, a baby cuckoo mimics the sound of an entire brood of hungry chicks. It is known that the volume of begging calls produced by a brood can influence the provisioning behaviour of the parents; hungry chicks call more and elicit a stronger response from their parents, which increase the number of feeding visits made to the nest. By taking the deceit to such a level, the baby cuckoo is maximising the amount of food that is delivered.

Although I have encountered cuckoo eggs in the nests of several different species, I have only seen their chicks in the nests of the reed warbler. When small and naked, the cuckoo chick can be identified by its dark-coloured skin and hollow back, the latter a sinister morphological feature that helps the chick to heave any host eggs or fellow nestlings over the rim of the nest cup and out into oblivion. Cuckoo eggs usually hatch a day or more ahead of the reed warbler eggs alongside which they have been incubated, and the resulting cuckoo chick soon gets to work removing the competition. As the cuckoo chick grows – the begging calls driving home its desire for food – so it occupies more and more of the nest cup until it seems to overspill the space available. It is the picture of gluttony. The spoilt, single child worshipped by its parents but born of deceit, an intentional orphan. That the cuckoo, and indeed many dozens of other species, indulge in brood parasitism – as this behaviour is known – underlines the evolutionary benefits on offer. The behaviour is not without its challenges however, not least the evolutionary arms race that has developed between host and brood parasite. There is a selection pressure on the hosts to spot the deceit and to avoid losing their reproductive opportunity, and some hosts seem better able to respond than others. While reed warblers will abandon a nesting attempt if they sense that their clutch has been parasitised by a cuckoo, dunnocks seem oblivious even though the cuckoo's egg is very different in colour and pattern from their own. Such differences in response reflect how recently, in evolutionary terms, the dunnock has become host to a brood parasite; the species has yet to develop a response. Fortunately, or unfortunately depending upon how you look at it, nest parasitism of garden-nesting dunnocks is rare these days, a reflection of the levels of population decline seen in UK cuckoo populations and the increasingly urbanised nature of our own population.

* * *

It is mid-June and a great spotted woodpecker has arrived at the garden feeding station with a youngster in tow. The adult, in this case a male who can be recognised by the small patch of red on his hind crown, has brought the fledgling to the mesh peanut feeder. The youngster has pale red feathering under its tail – in adults this is bright red – and a red cap; this latter feature often confuses those new to birdwatching into thinking this is a lesser spotted woodpecker, a much smaller and now much rarer bird. By bringing the youngster to the garden, the adult is effectively showing it a reliable source of food. The young woodpecker watches the adult feeding on the peanuts and then joins it. At first it seems unsure, unsettled perhaps by the motion of the hanging feeder as it responds to the arrival of this second bird. It glances around and then takes a few tentative stabs at the peanuts held tight within the wire mesh. It is evident that the young bird has dislodged some peanut fragments with these strikes, a few crumbs left sitting on the top of the half-open bill as the bird stares around rather blankly. The two linger for just a short while and then disappear. I see them again over the coming days, before finally, a week or so later, the youngster visits unaccompanied. It stays longer this time but does not return again, at least not in its juvenile plumage. Through this behaviour the adult has introduced the youngster to some of the local feeding opportunities, opportunities that are likely to prove reliable as the chick faces the first few weeks of full independence. I wonder if the bird has brought just the one chick to our garden feeding station, or were different chicks introduced to the peanut feeder on different days?

The visits of a great spotted woodpecker to a garden are not always seen as a good thing by garden birdwatchers, particularly if one takes an interest in one of the hole-fronted nest boxes erected for nesting tits or sparrows. Great spotted woodpeckers are nest predators, known to break into occupied nest boxes to take the chicks contained within. These provide an additional and valuable source of protein for the woodpecker's own chicks and so are targeted during that period when the woodpeckers have young of their own to feed. Interestingly, some great spotted woodpeckers have been seen to make several visits to an occupied nest box prior to actually predating the young inside. It is almost as if they are trying to determine the size of the chicks in the box, presumably to get the best return on the considerable effort that they will need to expend breaking it open. The woodpecker, when visiting the box, may make some exploratory taps against its exterior, perhaps to test the quality of the wood or possibly to elicit a response from the chicks inside, whose calls may reveal its occupancy. While some woodpeckers simply enlarge the entrance hole to access the chicks, others make a new hole in the side, just above the nest cup. Once in, they may make several visits, returning again and again to empty the nest of its unfortunate contents. In addition to the young of other birds, great spotted woodpecker chicks are fed on a diet of invertebrates, with moth larvae seemingly of particular importance. Box-visiting behaviour is most often reported during late spring, when boxes are full of nesting birds, but it can happen at other times of the year.

Great spotted woodpeckers are rare visitors here, the nearest broadleaf woodland some way from the garden and with plenty of other bird feeders in between, I've no doubt. It serves to remind me that the type of birds you can attract to a garden is shaped to a large degree by the presence of natural habitats nearby. If you want to attract farmland species, like reed bunting and yellowhammer, then your garden really does need to be rural in nature.

If you want nuthatch and great spotted woodpecker, you need nearby woodland, and if you want something as rare as cirl bunting or crested tit you need, as we have seen already, to be living in a small part of south Devon or the Highlands of Scotland respectively. Of course, most birds are mobile so there is always an outside chance of attracting something unusual; it is this tantalising prospect that adds to the thrill of garden birdwatching for many. I, for one, cling to the hope of an autumn afternoon when a wryneck drops in to spend a few hours on my lawn. This diminutive member of the woodpecker family was once a common breeding bird across southern Britain; however, following decades of decline, the wryneck is now a scarce passage visitor, seen in small numbers during spring and autumn.

As we discovered at the start of this book, features in and around the garden also play their part in shaping the community of birds that come to use it. The presence of some taller trees, either within the garden or on its boundary, can be a big help, as many thrushes and finches like to perch in these before dropping down to a feeding station. The presence of ornamental or native conifers can make a garden more attractive to visiting goldcrests or nesting greenfinches, while an area of lawn can prove attractive to starlings and other species that feed on soil-dwelling invertebrates taken from short turf. The two big features within the garden, however, are food and water, the latter especially attractive during the dry summer months. As we've noted elsewhere in this book, garden birdwatching benefits from an understanding of the birds that you hope to attract, their requirements and behaviours. Such knowledge helps to guide you in the choice of plants, foods and other features that you offer within the garden, ultimately increasing your chances of attracting a broad range of species throughout the year.

The tall conifers that border a garden a couple of doors away almost certainly hold the delicate nest of a goldcrest; the cycling refrain of the bird's song is evident most days now. The frequency at which the goldcrest's song is pitched means that some people, particularly those of more advanced years, can struggle to pick it out. This, coupled with the bird's small size and tendency to feed high in the canopy, may be one reason why the goldcrest is easily overlooked as a garden visitor. The goldcrest can be remarkably productive despite its small size, delivering two broods of six to eight chicks over the course of a season. The nest, which is a delicate cup of moss held together with spider webs and decorated with lichen, is usually attached to the underside of a branch and can be positioned anywhere between one and twelve metres above the ground. The potential productivity of this small bird is important, enabling its populations to recover from the impacts of cold winters, when significant numbers may be lost to the inclement weather. Male goldcrests can be particularly excitable birds, displaying aggressively at other males in a bid to hold onto their breeding territory. During these displays, there is much calling and posturing; individuals lower their wings and raise the orange-red crest that gives this bird its name. The central flame-red feathering of the crown is bordered with yellow and then two broad brushstrokes of black. A male presents the crest to his rival, so it has a clear signalling function and presumably imparts information as to his status. Outside of such disputes, the flame-red crown is hidden by other feathering and is far less obvious to the human observer. In the female, the central crown feathers are lemon-yellow.

After a few days of watching I have a sense of where the nest is located, even though I cannot make out its outline or see its thick mossy walls. The birds keep returning to the same bit of the nearest conifer, working their way up one branch and then disappearing into the dark underside of another. The visits are intermittent at first but later become more regular, and there are times when both birds arrive within a few moments of each other. This suggests that they are now feeding chicks, something later confirmed when I spot one of the birds leaving with a white, roughly pea-sized object in its bill. I am fairly certain that this is a faecal sac, the droppings produced by a chick and held together, encased within a globular skin. A chick about to pass a faecal sac will present its rear end to its parent, which then takes the sac as it emerges before carrying it away from the nest. This helps with nest sanitation and reduces the amount of white splashing which might give away the location of the nest to potential predators. Parent birds often swallow the faecal sacs produced by very young chicks, but those of larger chicks are invariably carried off in the bill and dropped away from the nest. In some species, notably the finches, the adults appear to stop removing the faecal sacs during the last few days that the chicks remain in the nest. Whether this is a change in the adult's behaviour or a change in the nature of the faeces produced by the chicks is unclear, but the result is that the rim of the nest cup collects an increasing number of droppings, giving it a white and crusty appearance.

* * *

Garden living is not without its challenges. In addition to the presence of cats, netting, windows and other agents of mortality, many small birds can struggle to find sufficient insects and spiders with which to feed their growing young. The urban environment can also prove challenging when it comes to the production of a clutch of eggs. One of the most important resources for a bird about to produce and lay a clutch of eggs is the calcium required to form the eggshells. A female blue tit will need to find 0.25 grams of calcium in order to produce her clutch of eight to ten eggs. Given that she has just 0.18 grams in her 0.5-gram skeleton, the bulk of this calcium has to be sourced from the environment and collected during the period over which the eggs are being formed and laid. Faced with the challenge of finding sufficient calcium, a female great tit may spend just under half of the available daylight hours searching. Female house sparrows, and probably many other garden birds, source this calcium from snail shells collected through diligent searching. Female spotted flycatchers, on the other hand, have been found to target woodlice, whose calcium-rich bodies provide an alternative source of this precious commodity.

It is thought that there have been long-term declines in the availability of calcium in the wider environment, associated with increased levels of acid deposition because of human activities since the Industrial Revolution. It has been found, for example, that undisturbed soil samples collected in southern England and then stored, show declining levels of soil pH over time. There is also evidence that the thickness of blackbird and thrush eggshells has declined over time. Rhys Green, Honorary Professor of Conservation Science at the University of

Cambridge, who has looked at this, used egg collections housed in museums and taken at known dates in the past. His measurements of eggshell thickness have revealed widespread declines in thickness since the nineteenth century. Such findings give cause for concern, since thin-shelled eggs are more likely to be damaged during incubation than those that are thicker. Recently, it has been discovered that long-term changes in the patterning on the eggs of breeding great tits may also be linked to eggshell thinning. Although eggshell patterning has several different functions, it can also have a structural role, with the deposition of the pigment protoporphyrin linked to eggshell thickness – in great tits more of the pigment is added to thinner areas of the shell, providing additional strength. The patterning change evident in a long-term series of great tit egg data correlates with a 6.5 per cent decline in average eggshell thickness and a matching decline in soil calcium levels at the study site, delivering compelling evidence for a link between the two.

At this point it is also worth noting the variation in egg size that is often seen between or within a clutch of eggs, something that I see fairly regularly in the great tit and blue tit nest boxes that I monitor each year in a local woodland. Looking across the studies that have been carried out, it appears that egg size is generally smaller in larger clutches, that it tends to be smaller in those clutches produced early in the season and in those laid at sites where the density of the species concerned is high. The presence of a 'runt egg', much smaller than the rest of the clutch, is often noted by those monitoring nests for BTO's Nest Record Scheme. Such variation suggests that resource availability can influence egg size, something that is important because the size of an egg has consequences for the emerging chick later in its life. Hatching success, fledging success and fledging weight are all influenced by egg size, the latter of particular significance because fledging weight is strongly linked to the chances of an individual recruiting into the breeding population the following year. If you are a chick hatched from a small egg, then the chances of you recruiting into the breeding population have already been reduced, such is the link between egg size and future prospects.

While gardening in a wildlife-friendly way can help to increase the numbers of invertebrates in a garden – including the snails and woodlice that provide nesting birds with a valuable source of calcium – it is also possible to provide calcium more directly, in the form of oystershell grit. This is added by some bird food companies to their seed mixes and it is certainly worth checking to see if the seed mixes that you use in your bird feeders during the spring and summer contain this useful ingredient. If it has been added to the food, then the chances are that the supplier will want to promote this in some way on the packaging.

Another incredibly important resource, required for the successful production of a brood of chicks, is an abundance of invertebrate prey. Most small birds feed their chicks on insects, spiders and other invertebrates, and they need to find significant numbers of these over the two-week or so period that the chicks are in the nest. It is well known that birds breeding in gardens tend to produce fewer chicks than those breeding in natural habitats, like deciduous woodland, and that those chicks reared in gardens tend to fledge at a lower weight than those raised elsewhere. This, coupled with the increased levels of chick starvation in gardens, suggests that the availability of invertebrate food is much lower here than that found in a piece of oak-rich woodland, for example. Studies looking in detail at diet show that woodland chicks receive a better-quality diet than garden-reared chicks, receiving a much greater proportion of caterpillars, and fewer spiders, beetles and adult flies. Watching a pair of blue tits foraging in the ivy-covered hedge, I am conscious of these studies and the increased challenges that these birds face breeding in my garden, rather than in nearby woodland. Having said this, as we shall see later in the book, there are benefits to garden living.

Research carried out in urban Leicester, and looking at house sparrows breeding in nest boxes, found that one in four of the nesting attempts failed to fledge any chicks. High levels of chick mortality were behind this, with most of this mortality occurring within the first four days after hatching. Chick survival was found to be strongly linked to diet, with those chicks receiving more invertebrates doing better on average than those receiving a greater proportion of plant material. Such findings demonstrate very strongly that gardening in a wildlife-friendly manner to boost invertebrate numbers is one of the most positive things that you can do to help nesting birds in your own garden. The plants in your garden will shape the community of insects and other invertebrates present, as will the extent to which you utilise insecticides and other chemical products.

Your garden flora is likely to be made up of a mixture of both native and non-native plants; some of the latter will be unsuitable for native insects because of structural characteristics or their chemical composition. A good example of this is the double cultivars of familiar garden plants popular with many gardeners because of their unusual appearance, more showy blooms or longer flowering season. However, in many cases these double cultivars typically carry less nectar, produce less seed or have structures that prevent access to pollinating or nectar-feeding insects. For example, the double cultivar of bird's-foot trefoil *Plenus* does not produce any nectar or pollen, so is useless for insects that feed on either of these resources. While native species are often the best plants to use in your garden, many non-natives can be equally valuable, especially where they can extend the flowering season for nectar-feeding insects; something that is particularly useful early in the year. If you can use plants to encourage and increase the numbers and variety of invertebrates in your garden, then this will mean there is more invertebrate food available for the chicks of garden-nesting birds.

* * *

With a shift in the weather fronts we experience several days of rain and a drop in temperature. These conditions, very different from those of the previous week, will make things more difficult for nesting birds and the insects on which they depend. It is not just the birds feeding within the garden that will experience the challenges of finding food; such conditions may also have an impact on the swifts whose early nests may already contain young. Wet and windy conditions make foraging difficult for these aerial feeders; with fewer flying insects on the wing, the adult swifts are unable to bring as many meals to the nest. This fall in food deliveries can greatly increase the length of time that the young swifts remain in the nest, extending the nestling period from perhaps as little as thirty-five days to as many as fifty-six. It is not just that the chicks may develop more slowly because they are receiving less food; if things are particularly bad then they may enter torpor, their metabolism slowing down and their energy demands reduced in an attempt to ride out the challenging conditions.

To some extent the adult swifts can counter local weather conditions – assuming the rain is just a local problem – by foraging much farther afield. A colleague of mine, interested in the foraging behaviour of breeding swifts, fitted a number of breeding birds with miniature tracking devices at a breeding colony on the East Anglian coast. The results that emerged were fascinating, revealing that some of the adults foraged many miles inland and away from the colony, taking advantage of the mass of insects associated with a crop of oil-seed rape. The birds had clearly been exploring the feeding opportunities over a large area, before finding this abundance of prey. Adult swifts can be very mobile in bad weather, their foraging trips covering perhaps as much as 2,000 kilometres. These predominantly non-breeding birds – pairs with young in the nest have to forage much closer to home – appear to head into the wind and skirt around the edge of a depression in search of better weather and improved feeding conditions elsewhere. Such movements may take our swifts across the Channel and onto the Continent.

Earlier in the book, we alluded to the fact that a typical swift colony will contain a mix of breeding birds and many non-breeding individuals, the latter not yet of breeding age. Unusually for such a small bird, a swift will not reach maturity until it is three or four years of age. At least some first-year birds return to UK breeding colonies the following summer, but it is thought that many others summer farther south, with more summering in the UK in the following years. As the tracking work above demonstrates, large numbers of swifts can gather where the weather conditions deliver an abundance of aerial prey. For example, they are often attracted to the concentrations of insects that can form near the coast, where sea-breeze fronts carry insects up into the air, a similar pattern also seen ahead of approaching thunderstorms. The timing of the swifts' arrivals and departures at UK colonies reflects the broader seasonal availability of invertebrate prey; they are one of our last summer migrants to arrive each spring and one of the first to depart, with the first birds leaving as early as mid-July. I do not think that they will be quite so early leaving this summer, the rain prolonging the breeding season as chicks develop more slowly. Food availability can also influence the number of eggs that a swift will lay, with early clutches typically containing three eggs, while those laid later usually containing just two.

Wet weather can also make things difficult for our other garden birds. Rain washes insects from leaves and other vegetation, reducing the numbers available to small birds, like tits and warblers, that feed in the canopy and seek out leaf-feeding caterpillars and beetles. The wet conditions can also leave adult birds with wet plumage, the moisture inadvertently brought back to the nest where it can chill nestlings and lead to increased levels of chick mortality. While chicks being reared within the relative comfort of a nest box may avoid the worst of the weather, heavy rain can be a particular challenge to those nestlings sheltering in open nests placed within garden bushes and shrubs. It is not just heavy rain that can be an issue; persistent light rain, extending over several hours or days, can be just as damaging.

A sense of freshness has settled over the garden following the recent rain, the vegetation lush and the ground soft underfoot. There is that delicious scent that comes after rain, a mix of earth and grass that is so pleasing. With so much moisture in the ground and vegetation, it is inevitable that each new morning brings with it a coating of dew. This remains through until mid-morning, later in the shade of the hedge, until the warmth of the sun causes it to evaporate. The soft ground means easier access to soil-dwelling invertebrates for the blackbirds and starlings, which focus so intently as they search for food. Feeding starlings favour the short grass that is characteristic of most garden lawns, each bird walking forwards and then pausing to probe the ground for leatherjackets (crane-fly larvae) and other soil-dwellers. The starling is aided in this by modifications to the skull and its associated musculature. These enable the starling to push its bill into the ground and then open it to create a hole. As the bill opens, so the bird rotates its eyes forwards, increasing the degree of binocular overlap and enabling the bird to better target its prey. An unexpected daytime hedgehog adds to the scene one morning, moving at speed along the gravel path before turning across the lawn and into the thick cover at the back of one of the borders. Tell-tale droppings on the lawn reveal that this is a regular visitor to the garden, though one we rarely see, a similar story to that of the visiting muntjac that somehow find a way into the garden at night.

The wet weather has benefited both the hedgehog and the blackbirds, increasing the numbers of slugs and snails active at night and seeing earthworms and other soil-dwelling invertebrates more active in the soil's uppermost layers – bringing them into the reach of a probing bill or even out onto the soil's surface, where a foraging hedgehog can also take advantage of them. A successful outcome for a garden blackbird nest may depend on there being sufficient rainfall to maintain access to these important invertebrate food sources. In contrast, extended periods of dry weather can be particularly damaging to both species, something that may become an increasing problem as our climate continues to change as a result of human activities. The rain has refilled the water butts and raised the level of the garden pond so that its shallow margins now hold water. The dried cases of recently emerged damselflies now sit closer to the water's surface on the water soldier, whose spiky narrow leaves dominate so much of the pond. The pond looks refreshed, the waterweed more green and the newts and water beetles more active; it has truly woken up over recent days and is looking at its best.

* * *

The final week of the month is a warm one, with clear skies of a uniform blue. I find that I am spending more and more time in the garden, tending to the vegetable beds and harvesting the first of the season's crops. There is something meditative in working the soil, a contemplative focus centred around a succession of tasks that are not too demanding, either physically or mentally. With sufficient warmth in the sun to be comforting, my senses wander to pick up the smells and sounds around me, even as my focus remains on the task at hand – weeding between the seedling vegetables. A rook call catches my attention, drawing my gaze up from the soil to the sky above. The bird's note is one of alarm and I soon spot the source of its agitation. A red kite is drifting across the garden on leisurely wingbeats, the warmth of this late morning hour sufficient to have plucked it from its roost and set it on its way. The rook, working in the company of two others, shadows the kite. The larger bird is not really a threat, at least not today, but with young rooks still in the nest, these adult birds are not taking any chances. This pattern of behaviour is repeated when the local buzzards put in an appearance, each one marked out through calls of alarm and then harassed and escorted on its way.

These are not the only birds of prey to drift across the garden, and I am no longer surprised by the numbers of larger birds more generally that cross the airspace above our small plot. The different birds come to my attention in different ways. Some, such as the oystercatchers that breed on the nearby farmland, are regular over the garden and alert me by their calls. In the case of the oystercatchers, their harsh, piping calls are heard throughout the spring and summer months, most often early or late in the day. Within a few moments of the calls being heard the birds appear, usually two birds together, passing above the garden on rapid wingbeats. Where the oystercatchers go in winter is unclear, but maybe they join others of their kind to forage on the nearby coast. Other birds, such as the regular sparrowhawk and occasional hobby, are revealed by the alarm calls of smaller birds. A snatch of the 'splee-plinck' alarm call from one of the local swallows and I know that a bird of prey is nearby. One group of birds that seemed particularly alert to the presence of larger birds of prey were the chickens that we used to keep. These could be relied upon to alert me, should something drift across the garden while I was head down, digging, planting or weeding.

I have only had swallows in one of the gardens that I have shared with birds and other wildlife. This was the most rural; an old plot that was once part of a fruit-farm established as a collective by an entrepreneurial back-to-the-land enthusiast more than a century ago. He had sold off small plots to a group of similarly motivated individuals, each working their own land but delivering a collective product for the city markets. Unfortunately, the plan had failed because of the location chosen. Set within the Norfolk Brecks, the site was subject to late frosts; these devastated the blossom, leading to poor fruit set and low yields. A handful of swallow pairs bred across the remaining land, now a series of large rural gardens and a working farm, and throughout the summer months the birds perched on the cables that crossed the garden from the nearby telegraph pole to the house. Their presence was a delight, not just because of their calls but also because they were seasonal companions present for a few brief months and then off. If I could have one bird added to the community with which I share the garden that I now own, then it would have to be the swallow.

Given the right conditions, nesting swallows will occupy less rural locations. I remember a pair that nested on a wall bracket in the alleyway between two houses in the very centre of a small Norfolk town. One of the houseowners must have found their presence unwelcome, because the following spring, just as the swallows had returned, a plastic bag full of polystyrene was tied to the bracket, removing the nesting opportunity altogether. There are some similar nesting opportunities on the edge of the town where I now live, the swallows this time very welcome and nesting around a small row of houses that face the common and a small pond. Our walks often take us along the front of this row of houses and it is wonderful to see the swallows, either perched on the wires or hawking low above the pond and the nearby long grass. It feels a much more welcoming place because of how the birds here are treated; they are welcome visitors, connecting these few houses with the distant lands within which they spend our winter.

Thoughts of swallows shift to those of swifts, the birds that are present nesting in the houses here around my garden. Glancing upward, I can see their tiny scimitar-like outlines high in the sky above, trawling for flies and other invertebrate life caught high on the wind. From here, the sky may seem empty of invertebrate life, at least once you get above the rooftops, but the presence of the swifts reminds me that there are significant quantities of invertebrate biomass up there. It is on these winds that aphids and moths, pollen beetles and tiny spiders riding their silken lines of gossamer, move vast distances, crossing counties and even countries. The thought of these journeys triggers a note of melancholy as I realise that, with June coming to its end, in just another five or six weeks these swifts will have gone. The year is passing her halfway point and the cycle of garden bird life is nearing a moment of change.

July

June slips into July and now that the bulk of the season's planting is done, work in the garden settles into an easier routine. There is still work to be done in the greenhouse, the tomato and cucumber plants now in their final places, and in the salad bed, with its succession of leaves and radishes. The wild lawn, which is left uncut throughout the spring and summer months, has been developing well and now hosts bright blooms and a wealth of insects. Grasshopper nymphs will soon complete their final moult before emerging into their adult form, and a myriad of different flies perch on the grass stalks until disturbed by a passing footfall. The dawn chorus has passed its spring peak, but there are still plenty of birds singing. Absent are those that make just a single breeding attempt each year, such as the blue tits and great tits, their attentions now directed towards the rabble of youngsters that jostle for position on the bird feeders. For the blackbirds, robins, dunnocks and song thrushes, the season continues and many pairs will now be on their second or even third nesting attempt of the year.

One species that typically only makes a single breeding attempt is the starling, and the first week of the month sees at least three youngsters fledge from the nest that has been built above the neighbour's front door. It is something of an unusual choice but testament to the starling's adaptability when it comes to selecting a suitable cavity in which to construct its rather untidy nest. Above the front door is an arch, recessed back into the brickwork and leaving a narrow ledge above the door itself. At some point in the past, a previous owner decided they didn't want a recess and so covered it with a piece of wood, cut to shape. Last winter this came loose, pulling away from the brickwork. Still not mended, it has created a giant nest box, the slit of an entrance on its upper margin formed where the panel has come away from the brickwork behind.

Over the past few weeks we have been able to watch the starlings come and go, ferrying in nesting material and, more recently, food. Early in the process the male would sing from the roof, sometimes using the television aerial as a song-post, sometimes the rim of the guttering. Male starlings sing throughout the year; outside of the breeding season starling song is thought to play a role in both signalling social status and flock formation. During the breeding season, however, it is very much associated with mate attraction and sexual behaviour. Early in the season, song production in male starlings is linked to possession of a suitable nest cavity, but after a male pairs with a female there is a general decrease in the amount of song produced. This change in the amount of song produced, coupled with the link to the ownership of a suitable nest site, underlines the song's role in mate attraction. Interestingly, male starlings deliver high volumes of song immediately prior to copulation, both with their mate and on those occasions where they indulge in an extra-pair copulation with the female from a neighbouring pair.

The song of this particular male has clearly worked its charms, as the young starlings fledging from the nest testify. The young starlings look very different to their parents. Instead of the glossy black plumage, with its 'petrol-on-water' kaleidoscope of purples and greens, the youngsters are rather plain, an almost uniformly pale grey-brown but slightly paler on the throat and slightly darker on the wings. Newly out of the nest, the young join their parents to feed on the short sward of the neighbour's front lawn. The chicks are noisy and persistent in the harassment of their parents, demanding food with vigorous calls. The family party disappears after just a few days – young starlings appear to have just a short period of parental care once they leave the nest – but it is impossible to determine how they have fared. We see few young starlings in and around the garden and I wonder if they have joined the growing flock of birds feeding on the fields that border the common. Over the coming weeks, the young will start to moult through into adult plumage, starting with their body plumage. For a few brief weeks, young starlings in various states of moult are present in the local flock, delivering a diversity of looks, the most comical of which are those individuals with the body plumage of a non-breeding adult – glossy black and speckled with soft brown flecks – but the plain brown head of a young bird. It looks as if someone has taken the head off one bird and stuck it onto the body of another.

* * *

While the breeding season may be over for this pair of starlings, other birds still have chicks to rear or even eggs to lay. Earlier on, we saw how important calcium was for egg formation, but clearly birds also need other nutrients if they are to produce a clutch of eggs. In addition to calcium, a female bird needs to secure sufficient protein, fat and water in order to make a viable egg. While the 1.3 grams of water needed for a great tit egg may not prove too much of a challenge to source, the protein may be more difficult to acquire, as may the fat. Fortunately, female birds can store small amounts of fat in preparation for egg production, perhaps laying down as much as a third of that required for a full clutch. If a small bird finds it difficult to source the necessary nutrients, then it can deploy a behavioural response. This could involve something as simple as increasing the interval over which the eggs are laid, something that has been documented in studies of breeding swifts, which increase the interval between successive eggs from two days to three.

Alternatively, and this is often seen in small birds, the male can provision his mate with additional food. Termed 'courtship feeding' by researchers, this behaviour is most often seen just before and during egg-laying, underlining its importance in supporting the female to secure sufficient resources for her clutch. The behaviour also serves two other functions: it helps to cement the pair bond and it provides the female with information on her mate's ability to source food. This latter point is important both in the context of the food the male provides to the female herself and in relation to his ability to provide for the nestlings once the eggs have hatched. While courtship feeding can easily go unnoticed in many garden bird species, often taking place at or near the nest, it can sometimes be seen in birds like the robin and woodpigeon, out in the open. Courtship feeding is most obvious to those who monitor the nests of barn owls and tawny owls for one of the national schemes. In both of the species, the male will bring small mammals and other prey to the nest site, forming a cache that the female can use throughout the egg-laying and incubation period. On some occasions I have visited owl nest sites and witnessed a dozen or more small mammal carcasses untidily stacked in the corner of a box. The presence of a cache, and its size, can provide a good indication of whether or not it is going to be a good breeding season for the owls.

The amount of resources needed to produce the clutch of eggs is determined by both the size of the clutch and that of each individual egg. The clutch sizes of garden birds tend to vary from two eggs – almost always the number seen in a woodpigeon or collared dove clutch – up to ten, the latter characteristic of a large blue tit or coal tit clutch. Clutch size tends to be fairly consistent within a species, but it can vary depending on the timing of the breeding attempt and the habitat in which it is made. The typical clutch size for most species is smaller in garden nests than in those made in other habitats – again suggesting the reduced resource availability in gardens compared to other habitats – and it also tends to be smaller later into the breeding season than it is towards the beginning, at least for single-brooded species. It is a little more complex with those species making multiple breeding attempts, such as the blackbird, where clutch size tends to peak in the middle of the season. This is something that I have witnessed through my nest monitoring work, where the first blackbird nests of the year (those recorded in late March or early April) typically contain three eggs, while those recorded in May or June invariably contain four.

Since the production of each egg represents a significant investment on the part of the female, it is likely that the number of eggs in a clutch will be shaped by the number of resulting chicks that the pair can support through to fledging. Produce too many eggs and you may find that you have more mouths to feed than you can support, and you increase the risk of losing them all if there is insufficient food available with which to feed them. There is, therefore, a selection pressure on female birds to produce a particular number of eggs under particular environmental conditions. As we have already seen, egg size can also vary both within a species and within an individual nesting attempt. Once the clutch has been completed or, more typically, is just about to be completed, the female will begin incubation, as we saw back in May.

* * *

It has been hot and dry of late, reducing the amount of slug and snail damage on the delicate salad crop in the vegetable garden, but no doubt making things difficult for those birds with young chicks in the nest and dependent on the regular delivery of insects and spiders. The ground is incredibly dry; this is very evident when I dig over a new bed, and the earthworms and other soil-dwelling invertebrates must have moved down within the soil profile in search of moisture. A similar thing has happened in the compost heap, as I discover when I go in search of earthworms for a hedgehog that has put in a daylight appearance in the garden. For hedgehog, blackbird and song thrush alike, this must be a challenging time.

We tend to think of the song thrush as a predator of snails, smashing open their shells on paving slabs or large stones, used as an anvil, to reach the snail within. This behaviour is largely seasonal and is used by the song thrush to tackle periods of drought in late summer, when more favoured prey become less accessible. Once a snail has been found and forcibly removed from its shell, it is wiped on the ground and then eaten. Although blackbirds have been seen to take and deal with snails in a similar manner, their efforts are often clumsy in comparison and it is clear that this is a behaviour they do not often use. Even snails can become hard to find when it is very dry, retreating farther into the dampest corners of the garden to aestivate, a behavioural response to the tough conditions and one that seeks to reduce water loss. Both slugs and snails are very susceptible to water loss. It seems that spiders become particularly important to thrushes at such times, though they are clearly not the preferred prey of these birds.

The run of hot days, coming off the back of the long dry spell, also sees an upturn in activity at the garden pond. Throughout the day, the shallow end of the pond receives a succession of avian visitors, coming to drink and bathe. Most arrive on the fence, scanning carefully before dropping down into the shrubs that crowd the corner of the garden behind the pond. Blackbird, starling, goldfinch and many others work their way through the shrubs to reach the water's edge. Many perch on small branches placed in the pond to provide easy access to the water, while some of the larger species – notably blackbird and starling – drop straight onto the densely packed waterweed to bathe. Sitting quietly on the other side of the lawn, I watch as they duck their heads down and flick their wings to lift a shower of water onto their plumage. One afternoon, I set up the camera next to the pond and film a blackbird doing this; slowing the resulting video down on the computer, I can see the different actions that facilitate this daily routine, which are needed to keep the plumage in good condition. That same afternoon, I see a pair of blackcaps arrive to drink, confirming the suspicion that they are breeding nearby. Over the coming days, I see them several more times, always arriving from the same direction and always leaving the same way. Almost all of the visiting birds drink in the same way, scooping water into the open beak and then tilting their head back to allow the water to run down with gravity. Pigeons differ in having the ability to suck water up and swallow it without the need to tilt their head back. This enables them to take on water more quickly, meaning they have to spend less time at the pond, which presumably reduces their exposure to potential predators.

One of the local blackbirds has discovered a very different opportunity that is being provided by the pond. Standing in the shallows, it strikes out to catch tiny froglets, which it then bashes against the ground before swallowing. This is a behaviour that I have seen before at a previous garden pond: that time the blackbird was taking tadpoles and, once, an adult smooth newt. Sad to witness – I have always thought that a pond is not complete without either frogs or newts – though it does underline the opportunistic and resourceful nature of birds. Is a tadpole or a newt so different from an earthworm or large beetle? Seemingly not, in the eye of a hungry blackbird.

Not all of the birds visiting the pond use it to bathe, or indeed as a source of amphibian protein. The house sparrows visit to drink but prefer to 'bathe' in the loose soil of the neighbouring flowerbed. Several dust-bathe alongside one another, the dry conditions having created something of a dust bowl in an area that is currently free from vegetation. Working the dry soil in among the feathers helps to clear out feather parasites in much the same way as a water bath would do. Seeing the house sparrows using the soil in this way reminds me of the robin that I once witnessed, sitting on some bare soil, with its wing and tail feathers spread wide. Although not dust-bathing, the robin was almost certainly using the sun's warmth to drive parasites out from between its feathers, a behaviour more commonly seen in blackbirds. Sun-bathing in this way can also help the bird's preen oil to spread across the feathers. Preen oil, which is produced from a gland just above the base of the tail and towards the bottom of the bird's back, has antimicrobial properties which inhibit the growth of bacteria that can degrade feather quality, and insecticidal qualities that work against feather lice. Watch a preening bird and you will often see it work its head back to the base of the tail and the preen gland, before working carefully through a section of its plumage, perhaps the breast feathers or those of the wing.

Although we do not get green woodpeckers visiting the garden here, this is another species that may be seen indulging in a bit of feather care, in this case through a behaviour referred to as 'anting'. The behaviour involves the bird positioning itself on top of an active ant nest, provoking an aggressive response from the ants, which spray the bird with formic acid. It has long been thought that the formic acid helps to control feather parasites and limit damaging bacteria, and it is certainly the case that birds like the green woodpecker and American blue jay show a strong preference for those ant species that produce formic acid over those that do not. However, the quantities of formic acid reaching the plumage are insufficient to have any real effect on the feather-damaging bacteria, though they might have more of an effect on the feather mites. It is possible that anting behaviour may serve another purpose, however, the aggressive response prompted by the bird's behaviour effectively emptying the formic acid reserves carried by the ants. This is beneficial for a bird that is going to go on to eat the ants, because high concentrations of formic acid in the stomach can prove deadly, especially if a bird is foraging on an empty stomach. Whatever the ultimate reason for this behaviour, it is fascinating to watch and something that I would love to witness again in my garden.

Woodpeckers nest in cavities, excavated from the decaying heartwood of a living, dead or dying tree, and this is one reason why they only tend to be seen in those gardens located fairly close to an area of suitable woodland or a farmland hedgerow with mature deciduous trees. This is why, more broadly, cavity-nesting species tend to be less common in urbanised landscapes lacking the mature trees and natural cavities required. Of course, we have made up for the lack of natural cavities to some extent, through the provision of garden nest boxes, particularly the hole-fronted nest boxes popular with nesting blue tits and great tits. Unfortunately, it is almost impossible to attract woodpeckers to use a nest box, a reflection of their nest site requirements which see them prefer to excavate each nest hole afresh rather than occupy one that has already been formed.

Unlike our other woodpeckers, which drum on resonant trunks and other surfaces to alert other individuals to their breeding territory, green woodpeckers rarely drum. This may have something to do with the fact that the green woodpecker's bill is weaker than that of our other woodpecker species, only really being used to chisel at soft decaying wood, and it may also explain why the species is more vocal, its loud, far-carrying yaffle call a familiar sound to those who live alongside these birds. Green woodpeckers seem more at home on the ground too, perhaps underlining that they take less invertebrate food from within timber and specialise instead on ants. Because of this, they only rarely visit garden feeding stations, but they have been recorded taking fat, mealworms and fruit from bird tables and ground feeding stations. In the winter, as we have seen, they can be attracted to windfall apples.

The dry spell finally breaks and a run of thundery showers delivers us from the hot nights and dusty air. The lawn and borders need the rain and there is a sense of relief at its arrival. So much rain falls that puddles form on the garden path, attracting the local woodpigeons who settle into them for a bathe. The birds adopt an unusual posture, raising first one wing and then the other to allow the rain to wet the plumage of their underwing. Once sufficiently wet, though not sodden, the pigeons shake themselves, preen and depart. I wonder whether they get any sense of pleasure from this behaviour, any recognition of being able to bathe properly after weeks of drought. While the rain is needed, I do wonder about its impact on those birds that still have active nests and young chicks. Some may be vulnerable because the nest is exposed, placed just below the canopy in a bramble or nettle bed, while others, safe in a nest box or cavity, may become chilled as parents visiting with food arrive wet. I worry more about nests away from the garden, especially the willow warbler and chiffchaff nests that I am monitoring at a nearby site; both can be quickly saturated by heavy rain, and I worry what I will find on my next visit.

* * *

The cooler, cloudier conditions that follow are accompanied by a greening of the lawn and a flush of new growth for some of the herbaceous perennials cut back earlier in the year. They also see the arrival of a family party of mistle thrushes, a brief and noisy visit from birds that more often I see in the local park or cemetery while out on a walk or run. The two adults are accompanied by three chicks, the latter noticeably paler and more strongly patterned than their parents. This family group is likely to remain together for a fortnight or more, ranging locally as the young move towards independence. These are active birds, foraging on the ground for grasshoppers, beetles and other insects, but despite their bold and upright appearance they are nervous visitors to the garden. The harsh alarm call, reminiscent of an old-fashioned football rattle, is soon uttered and the party departs.

These are not the only youngsters to arrive, the feeders busy with young tits and finches still recognisable as such by their plumage. A young robin is caught halfway between its juvenile and adult plumage, its breast feathering a mix of blood-red and brown as it moults through. The robin will replace its body feathers, together with some of those on the wing, a common pattern in small birds and something that we noted earlier in the year when the first of the young birds arrived at the feeding station. Once this young robin has acquired the red breast of an adult, it will be viewed as a rival by the local, territory-holding birds. The importance of the red breast is underlined by the way in which an adult robin will display at a rival. Where two robins are involved in a territorial dispute, the territory-holding bird will attempt to take a perch above the intruder. Once perched, the bird adopts a horizontal posture, displaying as much of the red breast as possible. If the territory owner finds himself below his rival, then he will throw his head back, again to emphasise the size of his breast. Displays of this kind provide an important function, enabling potential conflicts to be resolved without the birds coming to blows. For this approach to work, the 'badge' that is being displayed has to be an honest signal of an individual's quality, ensuring that a weaker rival can resolve that his chances of winning any physical encounter are so slight as to warrant the risk of physical damage not worthwhile.

Robins are aggressive birds and injuries are not uncommon where display of the red breast fails to resolve an encounter between two rivals. Although such physical encounters are usually preceded by display and aggressive vocalisations, they can kick off abruptly, the conflict escalating and the birds tumbling to the ground interlocked with one another and with each attempting to pin the other to the ground. Injury to the eyes or even death may be the outcome of such encounters, and this is one reason why territorial display has evolved as a means of resolving conflict. Because of this, it is hardly surprising that a young robin does not acquire the red breast straightaway. Prior to the adult plumage developing, the young bird can move through other robin territories largely unmolested, occasionally resorting to uttering chick-begging calls if approached by a particularly quarrelsome territory owner. It can play the 'I am chick and not a threat' card, tapping into those behavioural responses that would have been active in the adult when it had chicks of its own in the nest. Once it acquires the adult plumage, it will become a rival and no longer be tolerated.

It is not just robins that, despite their appearance, have another side to their character. The swallow is a species with a particularly sinister side to its character. As we have seen, this migrant visitor – so strongly associated with summer – may breed in a garden if it has a suitable structure. Car ports, porches and stables are used and some pairs may even nest in more urban sites. The nest, which is made from pellets of mud, strengthened through the addition of fibrous plant material, is usually placed on a ledge or beam, and it is this that becomes the scene for some rather unsporting behaviour. Unpaired male swallows may sometimes visit the nest of an established pair, sneaking in when the parent birds are away feeding elsewhere. During these visits, the unpaired male may drag young chicks from a nest to leave it empty. The failure of an active nesting attempt increases the chance that the established pair will break up. If this happens, then the unpaired male stands a chance of securing the divorced female for his own breeding attempt, her original mate shunned for his failure to deliver a brood of fledged chicks. The homeowner, aware of the swallow's breeding attempt, may blame its failure on the local cats – the dead chicks below the nest being evidence of the atrocity – but in reality it was an act of swallow infanticide. Such behaviour has no doubt evolved because it increases an individual's chances of seeing its genes passed on to the next generation, and when you have just a few breeding opportunities in your short life it makes sense to use any means to make the most of them.

* * *

A sick-looking greenfinch, fluffed up and reluctant to leave the feeder, gives cause for concern and I suspect that it is suffering from finch trichomonosis. Although the condition cannot be treated in the wild, I can help by removing my feeders, cleaning them and then holding off the provision of food for a few weeks in the hope that the birds will disperse elsewhere, reducing the chance of the disease being passed to other individuals through contaminated food. The fluffed-up condition and lethargic appearance show that the bird is unwell but do not point to the cause, since they are general symptoms of ill health. Back in the 1990s, the most likely cause of sickness would perhaps have been salmonellosis or colibacillosis, but now it is likely to be finch trichomonosis, a disease that first took hold in the greenfinch population in 2006/2007. Already known as a disease of pigeons, doves and some birds of prey, trichomonosis was first reported from UK finches in

2005, its presence revealed through regular monitoring and surveillance work. It is caused through infection with a protozoal parasite called *Trichomonas gallinae*. In finches, this tiny parasite infects the upper regions of the bird's alimentary tract, resulting in lesions that inhibit the bird's ability to swallow food and sometimes leading to secondary infection.

At the time that the disease emerged in UK finches, I was involved in a monitoring programme, the Garden Bird Health initiative (now called Garden Wildlife Health), which saw the systematic weekly recording of disease evidence across a range of garden bird species by a network of volunteers, paired with a number of veterinary scientists able to carry out post-mortems on some of the birds found dead. During 2006, our volunteers logged some 1,054 incidents, involving some 6,300 dead greenfinches and chaffinches (the two species most affected). Confirmation that *Trichomonas gallinae* was the disease agent came later, following DNA-sequencing and scanning electron microscopy. By looking at the disease data alongside data on greenfinch and chaffinch numbers, both within gardens and across the wider countryside, we were able to measure the scale of the impact of this disease outbreak. Our figures suggested that we had lost 35 per cent of our greenfinch population and 20 per cent of our chaffinch population in areas of high disease incidence over just a few months. This represents a significant impact, and one on which we have been careful to keep a watchful eye over the subsequent months and years.

The pattern of disease emergence was interesting and we could chart its movement across the UK and, later, into Scandinavia – the latter most likely the result of one or more infected chaffinches returning from UK wintering grounds to breeding sites in Sweden and Finland. The UK outbreak started in the West Midlands and southwest England; the following year it was centred on East Anglia and southeast England, and the year after that the centre of the outbreak shifted to Scotland. The DNA-sequencing work suggested that the origins of the disease lay in woodpigeon, a species that, as examination of hunted woodpigeons has revealed, can carry the parasite. Woodpigeon numbers have been increasing at garden feeding stations over recent decades because of changes in agricultural practices, and it is likely that this led to a spill-over event. UK greenfinch populations had been increasing prior to the emergence of the disease but, some 15 years on since the disease emerged, we have yet to see any evidence of a recovery. Although chaffinch populations took a hit at the time in those areas of high disease incidence, we did not see any real impact at the national level over the following years. However, since 2012 the chaffinch population has been in freefall and work is now urgently needed to determine whether finch trichomonosis is responsible for this sudden change in fortunes.

One of the questions about the disease that still remains is why was greenfinch so badly affected? Other garden bird species were found with the disease, including house sparrow and goldfinch, but none of these (with the possible exception of chaffinch) seems to have been hit in the same way. Is there something about greenfinch morphology or behaviour that makes it particularly susceptible? There is also an interesting question about the emergence of the disease in North America in 2007, there being no obvious migratory link between the finch populations here and the American goldfinch and purple finch populations that have been hardest hit in North America. One thing that the emergence of this disease has taught us is of the value of large-scale citizen science in disease surveillance.

Although currently underfunded, and living a somewhat hand-to-mouth existence, Garden Wildlife Health delivers a much-needed mechanism for identifying the arrival and emergence of other diseases. Some of these, such as West Nile virus, give particular cause for concern, not just because of their possible impacts on wild bird populations but also because some of them can then be transmitted to humans. West Nile virus, for example, can be transmitted to humans via mosquitoes; while rarely serious, severe infections require hospital treatment and in rare cases the virus can lead to meningitis. In autumn 2020, West Nile virus was confirmed in a whitethroat in the Netherlands, the bird seemingly having contracted the virus within the country – it had been captured earlier in the year and was at that time free of the disease. The virus has been found in mosquitoes inhabiting the same site as the whitethroat, so there is every chance that it will soon emerge here in the UK. Whitethroat is a summer visitor to both the Netherlands and the UK, and we have a sizeable breeding population.

The presence of a diseased bird in my garden also underlines the need to be careful with your own approach to hygiene around garden feeders. In the case of both salmonellosis and colibaccillosis, there is a risk of the disease being passed to humans via contaminated surfaces, such as bird tables and hanging feeders. I remember a letter I once received in which a lady proudly informed me that she washed her bird feeders in the kitchen sink, using her best dish cloth and tagging the task onto the end of her more usual washing up. This news sent a shudder through me, as it would anyone who has worked on bird disease! Bird feeders are best cleaned outside, in a bucket and with tools used only for that task. A pair of marigold gloves is helpful, followed by liberal use of soap and water once you have finished.

* * *

With the bird feeders removed, cleaned and stored in the shed, the second half of the month lacks the hustle and bustle that would normally be evident from the kitchen window. This pause in food provision comes just ahead of the seasonal switch from bird tables and feeders busy with young birds to a much quieter garden, a pattern that reflects the growing abundance of natural foods in the hedgerows and across the wider countryside. The garden is still busy, not least because of the pond. This continues to attract its succession of daily visitors, coming to drink and bathe. The thick hedge that runs up the side of the garden holds its chorus of house sparrows, and the flowerbeds and borders continue to be well used by ground-feeding blackbirds, robins and dunnocks.

Each passing day brings with it the growing sense that the swifts, *our* swifts, could leave any day now. One morning I will be outside, perhaps working in the garden or walking down through town, and be struck by a sudden absence. The moment when I realise that the swifts have gone brings with it the sense of the year's end, even though the dark days of winter are still months away. There is the realisation that the swifts will not be back, the summer will not be back, until the following May. From that moment on, a sense of loss permeates the remaining summer weeks; the season has been diminished by the swifts' departure, even though the house martins and swallows remain. How can a single species draw out such strong emotions, especially when it remains so distant from us during the short period that it is here? The swift interacts so briefly with us, an aerial wonder whose breeding attempts are hidden from our view, squeezed into the small spaces that we unwittingly create within the structure of our houses. It lives its life on the wing, feeding, sleeping and mating high above us, up in the margins of our perception. Yet despite all this, it continues to hold a power over us and its presence has become attached with meaning.

The swift's cultural resonance is almost spiritual, born of the sense of anticipation that permeates the spring weeks spent waiting for its arrival, and shaped by the sense of loss that follows its departure. Were it here all year round, the swift would no longer have the same pull on our emotions, the same power to stimulate the words of poets and writers. The swift provides a link to the 'other', to the other places that it visits on its vast annual journeys, and to the other world that is the air above us, which we cannot experience save within the confines of a plane or glider. Free from the earth, the swift rises above us, able to range where it wills, tethered only by those few brief weeks when it must settle to lay eggs and raise its chicks. Swifts are the summer in bird form, and it is their arrival and departure that punctuate the changing seasons in my calendar. Now that they have gone my mind is drawn towards autumn, even though the richness of summer has not quite reached its apex. The swelling vegetables, together with new foliage added to the shrubs and perennials, push back at any notion that the growing season is done.

August

One of the things that I have noticed over the summer months has been the way in which the house sparrows use our ivy-covered hedge for their social gatherings. These birds do not breed in the garden, nor it seems do they breed in the neighbouring properties, but instead they visit from houses that back onto the cemetery, a flight of seventy or so metres. The hedge appears to act as a roost site for a dozen or more birds, so it is a small roost by house sparrow standards, but it is hard to be certain because of the thick cover. The sparrows can be heard calling in an almost conversational manner, a behaviour that is referred to as 'social singing' and which appears to have a role in social cohesion, reinforcing relationships within the colony. Earlier in the book, we saw how social status in house sparrow populations is driven by the black bib of the adult males, supported by other behaviours that serve to reinforce a dominance hierarchy. Quite how social singing fits into this is unclear, but it seems to be most important outside of the breeding season, particularly early in the year.

The number of house sparrows using the garden reaches its peak in August, a pattern that is typical of those UK gardens fortunate enough to hold this declining species. Given that house sparrows are essentially sedentary in their habits, the adults remaining at their breeding colony from one season to the next and the chicks typically moving less than a kilometre when they disperse, this peak in numbers is likely to reflect the recruitment of young birds into the population rather than the arrival of birds from elsewhere. The dispersal of chicks from their natal sites to neighbouring colonies takes place from June through into August, so the birds using the garden are likely to be a mix of local adults and the youngsters from nearby colonies. The difference between the size of the house sparrow population in August, as revealed by the numbers in gardens recorded by the weekly BTO Garden BirdWatch, and that in April, when numbers in gardens are at their low point, can be used as a measure of how successful a breeding season it has been. This is one of very few bird species for which such a measure is likely to be meaningful; most birds are much less sedentary and there is considerably more movement between their populations.

Back in 2014, we used BTO Garden BirdWatch data to examine regional patterns in this measure of productivity, 'productivity' being the term used by scientists to describe the rate of production of new individuals in a population. Our measure covered the whole breeding season and the two or three breeding attempts that each house sparrow pair would have made. Although we failed to find any difference in productivity between rural and urban habitats, something which might have been expected because house sparrow populations in these two habitats have shown different patterns of decline, we did find some significant regional differences. These regional differences revealed that the highest productivity values for UK house sparrows were in Wales, with an average of 1.46 young produced per pair per season, while the lowest, at 1.29 young produced per pair per season, were in the east of England. Perhaps the most interesting pattern revealed by our work emerged when we compared these regional productivity values with data from the Breeding Bird Survey on regional house sparrow population trends. This revealed that those regions with the lowest productivity values were also the regions with the most negative population trends, while those with high productivity values had stable or increasing trends. Geographically, the pattern fell along a northwest/southeast split, with house sparrow populations in the south and east of the UK faring badly, but those in the north and west typically doing rather better.

Various reasons have been put forward to explain why house sparrow populations in the southeast of the UK are faring worse than those farther north and west, with a decline in invertebrate prey populations thought to be a particular problem. Although adult house sparrows feed mostly on plant material, their chicks are reared on a diet of small flies and other insects; it is the lack of these invertebrates that is likely to be behind the low numbers of chicks being produced in most house sparrow populations. One consequence of this has been the long-term decline in house sparrow numbers, and it seems that the days of large urban sparrow flocks, gathered in parks and other areas of urban greenspace, are now long gone. House sparrow populations in England have declined by 70 per cent since 1997 and the species is now included on the Birds of Conservation Concern Red List, a sobering statistic for a bird that was once so common.

House sparrow is not the only garden bird to appear on the list; also on the list are spotted flycatcher, song thrush and mistle thrush, together with garden-visiting farmland birds like yellowhammer and tree sparrow. Over the last fifty years, spotted flycatcher populations have fallen by 89 per cent, song thrush by 48 per cent and mistle thrush by 55 per cent. We touched on the spotted flycatcher's changing fortunes in an earlier chapter but it is worth just looking at the two thrushes, not least because most garden birdwatchers will be unaware of the declines evident in the populations of these two familiar garden birds. Song thrush populations have been in decline since 1970, with the bulk of the losses witnessed over the following two decades. Since then the population has recovered a little but it remains a long way from where it was fifty years ago. Research suggests that changes in the survival rates of young birds after leaving the nest have been sufficient to have caused the decline. The underlying reasons for this decline in survival rates are unclear but they are thought to be linked to changes in farming practices, particularly the drainage of land and, possibly, increased use of pesticides. A lack of information on mistle thrush breeding success, recruitment and survival means that we do not have any idea of the causes behind the decline seen in this species, which began in the late 1970s and which has continued to the present day, with only brief periods of stability or recovery before another fall. This lack of information underlines the importance of the large-scale citizen science monitoring that forms the bed-rock of conservation action here in the UK. Quite simply, we need more people out monitoring mistle thrushes through nest recording and the ringing of their young. This, of course, is made more difficult by the fact that mistle thrushes like to nest high up in trees, out of reach of most volunteers.

One other garden bird that appears on the Birds of Conservation Concern Red List is the starling, whose UK breeding populations have declined by 52 per cent since 1995. Starling populations have also been in decline more widely elsewhere across Europe. Fortunately, starling is a species for which we do have a great deal of information on different aspects of its life cycle, from data on its breeding ecology, to information on its movements and on the survival rates of both adult and young birds. These data have been used to model how changes in different components of the life cycle, such as the number of chicks produced or annual survival rates, might alter the population's size. If the models produce a similar pattern to that seen in the data collected on population size from monitoring schemes like the Breeding Bird Survey, then this can indicate a potential cause. In the case of the starling, it appears to be changes in the overwinter survival rates of young birds that is driving the decline. The ecological drivers that lie behind this reduction in survival rates are poorly understood, but changes in the management of pastural farmland – an important foraging habitat for starlings – are thought to be largely responsible.

As we have already seen, starlings feed on soil-dwelling invertebrates, with the larvae of crane flies a particular favourite. Changes in how pasture is managed, from increased stocking densities to the use of insecticide treatments, may have impacted populations of favoured invertebrates, reducing the food available to starlings. Another factor may be the changing climate, with increased temperatures and a shift in patterns of rainfall reducing the moisture content of soils and impacting on the soil's invertebrate fauna. It is all too easy to overlook the intricate links that exist between birds, their diet and climate, links that may be broken should conditions change. One such link can be seen in the seasonal shift in starling diet. In late summer, the amount of plant material taken by feeding starlings increases, while the contribution from invertebrates declines. This seasonal shift in diet, which reflects the seasonal pattern of food availability, is matched by a surprising change in the starling's anatomy. As the proportion of plant material taken increases, so the length of the starling's intestine increases. Plant material is more difficult to digest than invertebrate material, and the starling requires a longer intestine to do this. It might seem surprising that such substantial changes can occur in what we would traditionally regard as being fixed forms like the adult bird, but we have long known of similar changes evident in the major organs of migratory species. In many migrant birds, for example, the reproductive organs shrink in the run-up to migration – they are not needed at this time of the year and would add unnecessary weight – while the heart and flight muscles increase in size ahead of departure. The digestive system appears to be particularly dynamic in migrant birds, increasing in size during periods of refuelling but diminishing when the bird is about to make a migratory flight.

Another problem that starlings face is the loss of nest sites. This is a cavity-nesting species, which often breeds in cavities under roof tiles or within barge boards. As building regulations have changed, so these nesting opportunities have been lost as access under roof tiles is blocked with tiny plastic grills, and wooden barge boards are replaced with plastic that doesn't rot. The loss of nest sites is something that can be addressed by garden birdwatchers and homeowners. A starling nest box, with an entrance hole of no more than 45 millimetres, a floor area of approximately 310 square centimetres and deep in structure, can make a valuable addition to a property and provide these Red-listed birds with an opportunity that would otherwise not be there.

* * *

A period of very warm weather settles over much of the country and I limit my time working on the garden to those first hours of the morning. To be out later in the day risks the damaging effects of the sun's rays and would only add to the sense of tiredness that comes with the hot conditions. Lunchtime walks take me away from the garden and along the river, whose shaded woodland margins cast deep shadows over the myriad insects dancing just above the water's surface. Even here you can feel the heat of the day. Back at home the garden is at its best in the evening, just before dusk. Evening brings a stillness and sees the thermometer drop back to more tolerable levels. Sitting quietly, with a cold drink at hand, there is time to take in the garden. The scent from the borders hangs in the air, while the sounds of birds and insects provide a pleasant backdrop against which to unwind and shake out the troubles of the day.

There is something inherently meditative about late evening, that period marking the transition between day and night. Sitting still, my presence in the garden becomes part of the background and allows me to watch and listen to the birds that are now busy taking a final feed before going to roost. The same is true for the day-flying insects that will shortly find a sheltered spot to sit out the night. At the same time, night-flying insects begin to emerge and I watch a silver Y moth hover and feed from the stocks and other flowers that remain open in a nearby border. So named because of the conspicuous metallic silver 'Y' mark on its wing, this moth is a summer visitor to the UK, arriving in large numbers to breed and present from May through to September. Here in the garden, its caterpillars could just as easily be feeding on my climbing beans as on the nettles that grow in a large clump near the boundary of the property.

Attracted by the moths and other nocturnal insects, at least three different species of bat hawk over the garden. Just as the light is beginning to go, I can sit and watch pipistrelle bats as they criss-cross the fading sky above me. Through the use of a bat detector I have been able to listen to their echolocation calls, the calls also revealing the identities of both common pipistrelle and soprano pipistrelle feeding over the garden. One other bat, whose identity I have confirmed by both sight and sound, is the brown long-eared bat. Unlike the pipistrelles, this species feeds by gleaning prey from vegetation. The combination of its behaviour – hawking close to the hedge – and silhouette helps to confirm its identity. Other species of bat probably visit the garden, but I'd need to leave a more expensive type of bat detector out overnight to record the presence of these hidden visitors.

The transition to night is accompanied by a chorus of blackbirds, 'chooking' nervously in the gathering darkness. This call, which seems to signal nervous agitation, echoes out from across the garden and beyond, as if each bird is reacting to its neighbour. Perhaps remembering the closing lines of Edward Thomas's poem 'Adlestrop', I imagine this chorus rippling out across the whole of England. Farther away, but still audible on the slight breeze, I can hear the murmur of rooks at the rookery and a young tawny owl. Earlier in the year, I had strained my ears to try and catch the strident rasping call of a male corncrake. This individual had set up territory on the edge of the town, in an unmown field over which the swallows hawked, an unexpected visitor that was almost certainly the result of a reintroduction programme taking place a few miles away. Sadly, I had not managed to catch his calls on the wind, the distance just too great, but the evening's listening had brought other calls, linking the garden to the landscape beyond. Other birds were roosting within the garden itself, the doves on their nest, the house sparrows and blackbirds in the hedge. Each one would be waiting for the coming day, alert to potential predators but hopefully safe and secure on its chosen perch.

* * *

It feels as if we are now beyond the end of summer, the borders looking tired and the evenings continuing to draw in. The swifts have been gone for several weeks now and it is only a late-passage bird that serves as a fading echo to the evenings past, when the birds' black scimitar-like forms chased across the sky above the house. Attention shifts away from the garden to thoughts of the first autumn migrants, passing south on their way to their wintering grounds. Despite the autumnal feel and the sense that summer's activity has passed, the woodpigeon breeding season is still in full swing and we have a bird sitting on eggs in the top of our ivy-covered hedge. The woodpigeon breeding season is a long one, the species recorded nesting in every month of the year, though the main period for active nests extends from April through into mid-October, with a peak in activity from mid-July to late September. Our pair has been sitting on their nest for at least a fortnight, but it is difficult to say when the clutch of two eggs was laid because woodpigeons often pass the day loafing on potential nesting platforms, especially in the run-up to egg-laying.

Two eggs might seem like a small clutch for such a big bird but it is very much the standard for pigeons, seen across just about every pigeon and dove species globally. Only rarely do you encounter a woodpigeon nest with anything other than two eggs; if you do, then it is likely to have just a single, glossy white egg. On average, just one in five hundred nests will contain a clutch of three eggs. The eggs are large, about 40 millimetres long by 31 millimetres wide, and it seems that two eggs is the maximum number that can be incubated successfully – hatching success has been found to be lower in clutches of three eggs in other pigeon species. Both parents share incubation duties, each taking significant shifts, and this means that the eggs are only rarely left uncovered. Since the eggs are large, obvious and easily accessible, keeping them covered reduces the risk of them becoming a meal for a passing magpie, crow or grey squirrel.

Woodpigeon chicks are curious-looking creatures, the narrow head and prominent bill reminiscent of the dodo caricatures once seen in childhood books. The chicks do not get any more attractive as they age, at least not until they have well-developed plumage just prior to leaving the nest. During the first few days after hatching, the chicks are fed entirely on 'crop milk', a substance that is very similar to mammalian milk in its composition, being rich in proteins and lipids. The nature of the nutrition provided by the crop secretion is very similar to that provided to the chick in its egg, just before hatching. The 'milk' is produced from special cells in the adult's crop, which slough off from the crop lining and are then regurgitated. Pigeons are one of just a small number of birds that feed their chicks on crop milk, a feature shared with the greater flamingo and emperor penguin. After a few days the chick's diet changes, with the introduction of plant and animal material and a reduction in the amount of crop milk provided. Initially, the chicks will be brooded continuously, receiving food at roughly hourly intervals, but once they reach nine days of age they will be receiving just two feeding visits per day, typically in the morning and evening. It is going to be a little while before we will know if our woodpigeon eggs have hatched successfully; until then, our work in that part of the garden will be watched over by the sitting bird and his or her beady eye.

* * *

Part of our lawn is left uncut, the longer sward good for insects and for a range of plants that are unable to cope with a regime of regular mowing. These include dandelion, whose 'dandelion clocks' were no doubt a part of many childhoods. Dandelion is popular with bullfinches, those in my parents' garden well used by the small family parties that used to visit. I used to watch entranced as these little finches either hovered to pluck individual seeds from the seed head or perched on the stem, which invariably bent under their weight. Goldfinch is another bird that can be seen taking seeds from flowers growing in the lawn. The species has a preference for seeds that are yet to ripen fully but are instead still in the milky stage of development. The feeding preferences of goldfinches extend across a number of familiar garden plants, including lemon balm, greater knapweed, various thistles and teasel. Interestingly, the seed heads of teasel present something of a challenge because its seeds are held deep within the seed head, placed between spines. While the bill of a male goldfinch is sufficiently long and narrow to enable the bird to reach the teasel seed, the slightly shorter, but equally narrow, bill of the female cannot quite reach the seeds. If she is to do this, then she has first to bend the protective spines out of the way, a time-consuming process and the likely reason why you typically only see male goldfinches feeding on this plant. In case you are wondering how to determine goldfinch sex, the red above the eye pushes back slightly behind the eye in the male but not in the female.

If you cannot tolerate the thought of teasel or thistles in the garden, both of which can settle in well and establish quickly, then perhaps consider some of the other seed-bearing plants that are favoured by goldfinches and other birds. These include devil's-bit scabious and field scabious, lavender and the lemon balm and knapweed already mentioned. Of course, there is also sunflower, which not only produces a showy flower but also plenty of seed. If you have the space then also consider seed-producing trees, like alder, hornbeam and silver birch. These may be visited by a wider suite of finches, including siskin, lesser redpoll and hawfinch. The last of these is an uncommon garden visitor, more usually associated with scrubby woodland or the ornamental landscapes characteristic of National Trust properties – hence the tongue-in-cheek name used by some of 'National Trust Finch'. Hawfinch is perhaps the smartest small bird you'll encounter in a garden setting, its attractiveness enhanced by its relative scarcity, and it is definitely a red-letter day should one of these turn up at your garden feeding station.

Siskins and lesser redpolls are the more likely garden visitors, the former fairly common to many gardens during the second half of the winter, the latter an occasional winter visitor. As we have already seen, the presence of these finches at garden feeders depends on the relative availability of tree seed within the wider countryside. Many garden birdwatchers refer to the lesser redpoll by the name redpoll, dropping the 'lesser' or even unaware that it forms part of the name. Redpoll taxonomy has been something of a minefield in the past, and to some extent still is, with some identifiable 'races' either split into distinct species or lumped together. Currently, lesser redpoll *Acanthis cabaret*, common redpoll *Acanthis flammea* and Arctic redpoll *Acanthis hornemanni* are recognised on the official British list, maintained by the British Ornithologists' Union. Confusingly, the common redpoll is not the common redpoll species in the UK but is instead an uncommon winter visitor. It is the lesser redpoll that is the common species found breeding across the country. There was a suggestion that lesser redpolls were increasing their use of garden feeding stations during the 1980s, but with their wider countryside populations now significantly smaller than they were in the 1970s, we are unlikely to see them become established as regular garden visitors in the manner that goldfinches and siskins have done. This is a shame, since these are delightful little birds.

Lesser redpolls feed on very small seeds. Acrobatic in nature, they forage in trees – alder, silver birch and downy birch are favoured – taking the small seeds from their cones and only venturing to the ground when the remaining seed has mostly fallen from the tree. They will, however, also take small seeds from bird tables and hanging feeders, though usually only when visiting garden feeding stations in the company of other finches, notably siskins. Rather unassuming in its plumage, an adult lesser redpoll has a smudge of red on the forecrown, just above the small, pale-coloured bill. This is more strongly marked in the male than the female, but absent altogether in young birds. They have still not visited the garden here, but I used to have small numbers visiting my previous garden through the late autumn and winter months, where they mainly came to the niger feeder. Redpolls can be more regular visitors to those gardens in the north and west of Britain, located within the core breeding range and close to suitable breeding habitat, such as birch scrub.

* * *

One morning, in the second half of the month, I emerge from the house and flush a grey heron from a neighbouring roof. I have occasionally seen them from the garden, in flight over the town, but this is the first one to have come close to actually being in our garden. I know that we are not the only house in this part of town to have a pond, and I suspect that some of these other ponds contain ornamental fish, so the presence of the heron should not be too much of a surprise. This is a young bird, characterised by its plumage, which is more buff-coloured in its tones than would be the case in an adult. As it takes to the air I wonder if this is its first visit, or has it already discovered the easy pickings that some of the local garden ponds are likely to offer? We do not have fish in our pond, viewing them as likely to limit the fortunes of the breeding smooth newts. However, there is no reason why a heron wouldn't give our pond a go; a newt or frog would make a suitable snack, though would probably be more difficult to catch than an ornamental koi carp or goldfish. Our pond has plenty of cover, something that is often lacking in ornamental ponds whose owners want to be able to watch their fish. Cover within the pond provides added protection for fish, amphibians, water beetles, and the larvae of dragonflies and damselflies.

There is an air of patience about a hunting heron, the way in which it stands motionless staring intently into the water, poised ready to strike at an unwary fish or amphibian. It is often said that herons do not like feeding in the presence of other individuals of their kind, hence the popularity of ornamental plastic herons set alongside garden ponds as a deterrent, but I suspect that these are much less effective than their owners might hope. Grey herons nest colonially, early in the year and, more often than not, high in the tops of suitable trees. They may also come together in numbers to rest after feeding, something that you occasionally see in the wider countryside. They are also more versatile in their diet than you might suspect. In addition to fish and amphibians, grey herons regularly take insects, crustaceans, small mammals, reptiles and even small birds, such as young ducklings or rails. I remember seeing one catch and eat a stoat one New Year's Day morning out on the marsh at Cantley in Norfolk. It was a remarkable sight and I can only imagine that the stoat had unknowingly strayed within striking range of the stationary heron: one moment a hunter and the next an item of prey. The heron had surprisingly little difficulty in swallowing the unfortunate stoat.

Herons tend to feed during the day, favouring early morning and evening, so it is entirely possible that this bird has been visiting the garden before we are up and out. Our pond is located in a part of the garden that cannot be seen from the house, so even though we are early risers, a heron may visit unnoticed. Other species, such as rook and jay, are also early visitors to many a garden, their visits over long before most householders venture out. UK grey heron populations have declined by roughly a quarter over the last decade, though this decline should be viewed against a longer-term increase in their breeding population. The numbers of grey herons breeding in the UK has been monitored through the annual Heronries Census, operated by BTO since the late 1920s, and it has generated a remarkable series of interesting data. The data show that populations can decline rapidly following a cold winter, such as that of 1962/63; unable to access food because of snow and ice, the herons starve and die. The strong downturn evident between 2005 and 2013 might be linked to cold winter weather or, perhaps more likely, the early spring gales that were known to have dislodged tree-top nests and their contents. Despite keeping an eye out for the heron it does not reappear, which is good news as far as the newts are concerned.

Another bird that puts in a single, passing appearance during August is an osprey, seen drifting high above the garden early one afternoon. This will be a bird heading south as part of its annual migration, a journey that is likely to take it down through Spain and France, and then on into Africa, where it will winter, probably in West Africa. If it is a young bird, and I suspect this one was despite the poor view I managed to secure, then it is likely to remain on its wintering grounds for at least its first summer, returning north perhaps in its second year. Even then, it is unlikely to make its first breeding attempt until at least its third year.

* * *

The sighting of the osprey is noteworthy, and it reminds me to look up more often when out in the garden. Of course, my garden and its resources offer nothing to a bird like an osprey, its presence overhead purely a matter of chance. It does, however, connect me to the other landscapes that this bird will encounter on its journey south, and to the coastal wetlands in West Africa where it is likely to end up. For my more familiar garden birds, any journeys made are likely to be on a scale that is far less grand. While some, such as the blackcaps that have been visiting the pond to bathe and drink, may leave the UK to winter around the shores of the Mediterranean or more centrally within France or

Spain, others will play out their entire lives within just a few miles of this small patch.

Over the coming weeks, as late summer shifts into autumn, we will see the arrival of familiar species, such as blackbird, chaffinch and greenfinch, that come not from the local landscape but instead arrive from breeding grounds far beyond it. This arrival will go largely unnoticed, the individuals involved hidden within the resident population because they are no different in their appearance. It will not be until after that, perhaps as late as early November, that we will finally see the arrival of recognisable winter visitors, such as redwing and fieldfare, revealed as immigrants because they do not breed locally. Until that time, the garden bird community remains fixed in its composition, the birds here the same as those present earlier in the summer. What has changed is the arrival of a new generation of young birds entering the population and now, for the most part, independent and alone. Of course, there are still ongoing breeding attempts for multi-brooded species like blackbird, song thrush, woodpigeon and collared dove.

Most of this new generation of young birds will move away from the garden and the nests within which they were reared. This process, known as 'natal dispersal', ultimately leads them to the territories within which they will themselves breed. Dispersal is an important process, not least because it reduces the chances of an individual pairing with one of its parents or another close relative. There are significant genetic risks associated with inbreeding, and behaviours such as dispersal have evolved to reduce these. For many species the natal dispersal distance isn't that great, perhaps just the width of a few breeding territories, but enough to minimise the risks of inbreeding come the following breeding season. It doesn't make sense to disperse too far, because the chances are that a chick will have been raised in a suitable breeding habitat – the nest has been successful after all – and the farther it goes the lower the chances of finding equally suitable breeding habitat elsewhere. Because of climatic patterns and geology, suitable habitats and breeding conditions tend not to be randomly distributed but are instead clumped.

One other aspect of natal dispersal is worth a mention here, and that is the differences often evident between male and female chicks from the same nest. Typically, female chicks move farther than their male siblings during the natal dispersal period – a pattern that is the reverse of that usually seen in mammals. This difference between male and female chicks reflects the different costs and benefits of dispersal faced by the two sexes. These might include competition for breeding territories (more likely to be a factor for males than females), skewed sex ratios (territory-holding males may suffer higher levels of mortality than breeding females, leaving more vacant territories closer to the one from which a chick has dispersed) and differences in size between the two sexes (such as those seen in the sparrowhawk). In addition, there is some evidence to suggest that differential dispersal between the sexes may also help to reduce the risks of inbreeding. If male and female chicks typically move the same distance away from where they were raised, then there is an increased chance of them meeting and pairing than if they moved different distances.

An indication of the distances moved by dispersing chicks can be seen in the information collected by bird ringers through the ringing of young birds still in the nest. I know from my own ringing activities how young tits ringed in one nest box study can turn up to breed at other sites nearby, although it is worth noting that the encounter rates for such chicks are low simply because the young birds have so many other nesting opportunities beyond those offered at sites monitored by bird ringers. Of course, another factor is the naturally high levels of mortality experienced by small birds, through predation, disease and starvation. The chances of a small bird, like a blue tit, entering the breeding population the following year are low, such are the harsh realities of life. For those young birds that have gained their independence over recent weeks there are a lot of challenges ahead, even if they can take advantage of the natural food supplies on offer in hedgerows and gardens at this time of the year.

While some measure of just how good a breeding season has been can be gained from the ratio of young birds to adults in the catches made by bird ringers late in the season, it will not be until the following year that the impact of this breeding season – good or bad – will be felt within the population. With the challenges of winter ahead there is still much that can happen; perhaps a mild winter might take the edge off a poor breeding season through a reduction in overwinter mortality. Conversely, the longer-term benefits of a good breeding season might be wiped out by a particularly challenging winter, which sees significant levels of mortality. For now, however, it feels like this breeding season may have been a challenge for those species dependent on soil-dwelling invertebrates, while those species nesting early in the year and dependent on leaf-eating caterpillars have probably fared that much better. We will have to wait and see what the results from the national monitoring schemes reveal, when they are published later in the year. Until then, my thoughts turn to the late-season nesting attempts that continue to be made in the garden: to the woodpigeons in the hedge, the blackbirds behind the greenhouse and the collared dove on a ledge above next door's extractor fan. These are the birds that will hold my attention as the August bank holiday approaches and the month nears its end.

September

There are still a few house martin broods around locally as August slips into September. This is despite the fact that others of their kind have already begun their journey south, heading towards wintering grounds in Africa. A glance up at a line of six nests, placed under the eaves of a property located towards the bottom end of town, reveals at least two that are still active. House martin chicks peer from the nests and at each entrance there is a little cluster of softly rounded heads. The birds' bright eyes hold tiny crescents of sunlight, the world outside captured in a glance. Any day now and these young birds will take to the wing for the first time, the beginning of an incredible journey that will take them far from their place of birth.

The house martin is one of those species whose populations give cause for concern, but it is also a species for which we lack reliable data on its changing fortunes. The martins nest in loose colonies but these can disappear suddenly, even from sites that have been used consistently over a long run of years. New colonies can also spring up elsewhere just as unexpectedly, so it is very difficult to monitor the population and to be confident about how it has changed from one year to the next. What data we do have suggest that UK house martin populations declined by a fifth between 1995 and 2018. The pattern of population change has been somewhat uneven, with English and Welsh populations in decline, those in Scotland seemingly stable and those in Northern Ireland increasing. What is interesting about these patterns is that they are similar to those of the swallow, a closely related species, which might suggest that our aerial-feeding insectivorous birds are collectively facing similar environmental challenges, perhaps linked to food availability, climate or both.

Despite having a house that appears to be suitable for house martins, this is a species that I have never managed to attract. Stays at farmhouse bed and breakfasts, with active colonies in residence outside the bedroom window, only add to the disappointment of not having these delightful little birds in residence here. There is something special about watching house martins fly up to a nest cup just above your window and to witness their dexterity as they rebuild a damaged nest or deliver food to growing chicks. The nest itself is an amazing construction, especially when you realise that the bird has had to build this just by using its beak. Each nest is built from the bottom up and may contain in excess of 900 tiny pellets of mud, all of them sourced from a muddy puddle or the margins of a pond. The reliance on mud underlines the difficulties that can arise from prolonged periods of dry weather. If the martins are unable to find suitable material then this may weaken the nest's construction. Added to this are the risk of the nest being knocked down by a homeowner unhappy about the presence of these birds on their pristine house, and the relatively lower risk of the nest being usurped by the local house sparrows or predated by a visiting tawny owl, the latter species known to pull down nests with chicks during night-time hunting forays.

The loss of a nest cup at the end of the summer to an unhappy homeowner can have implications for the following breeding season. House martins preferentially re-use old nests if they are available, an approach that avoids the time-consuming task of building one from scratch. Research has revealed that house martin pairs able to re-use an old nest save themselves ten days on average, time that can instead be invested in the breeding attempt. Pairs nesting in previously used nests are more likely to be multi-brooded, having two or even three broods over the course of the summer, and are also more likely to be successful with their nesting attempt. There is a balance to be struck here, however. Most other birds make a new nest ahead of each nesting attempt, an act that counters the likely build-up of parasite populations (such as lice and fleas) associated with active nests. Many of these parasites can survive in a nest between seasons, so those house martin pairs re-using old nests may suffer more from nest parasites than pairs starting from scratch. The savings gained from not having to construct a new nest must outweigh any negative impacts from facing more parasites when re-using an old nest.

The house martin chicks look like they are waiting expectantly for an adult, returning with food. It is possible that these young birds may have already made their first flight, since fledged young often return to the nest cup – or an adjacent one – to roost over the days that follow the first successful departure. I suspect that such behaviour is more common earlier in the breeding season than it is now, the young from those initial nesting attempts under less pressure to begin their migration. For these chicks, there must be more urgency to leave the nest and join the mixed flocks of swallows and house martins now gathering on the telephone wires. Adult birds encourage their young to leave the nest by flying slowly past the entrance, all the while uttering the characteristic contact call. An adult will usually escort its chick on its maiden flight, which often sees the young bird end up on the ground or perched rather clumsily on a nearby structure, at which point the adult attempts to lure it back to the nest cup. All I can see of these chicks are their heads, so it is impossible to tell how close to fledging they are or whether they have already made their first flight. I know that they will face many challenges over the coming weeks and months, but I hope they will return to find the nest cups on these houses untouched and ready for next summer's breeding attempts.

* * *

The tawny owl is a bird that we hear fairly regularly from the house throughout the winter nights, but this month we have heard a young individual calling on several occasions. That it is a youngster is clear from its poorly formed vocalisations: the pitch of the call isn't right and the refrain is often truncated before its final flourish. The familiar 't-wit t-woo' used to characterise the tawny owl's call is actually a representation of a pair of calling birds, the 't-wit' a reference to the female's 'kee-wick' call and the 't-woo' the male's drawn-out hoot. A tawny owl pair will often duet, the male hooting and the female adding her contact call in the pauses between the male's refrain. Tawny owls are highly sedentary in their habits and remain together as a pair throughout the year and indeed from one year to the next. This strategy is useful, since it enables the birds to use their extensive knowledge of their territory to locate small mammals and other prey more effectively, a particularly valuable skill in those years when small mammal populations are low – something that happens on a roughly periodic cycle of four to five years.

During September, however, the tawny owl breeding season is yet to get going and this young bird has presumably just gained independence and is now dispersing from the territory in which it was raised. As it ventures farther afield, it will inevitably intrude on the breeding territories of established pairs, triggering a series of calls from the resident birds. These territorial calls impart information about the territory holders, and it is known that the owls can identify each other from the pattern and structure of each call. Tawny owl pairs learn to recognise the calls of their near neighbours and respond less strongly to their calls than they would those of a stranger, not heard before. The arrival of a young bird draws a prompt response, with both members of the pair answering with calls of their own. As evenings spent monitoring tawny owls have taught me, territory-holding birds actively move towards the intruder and will attack the bird if it refuses to move away.

The tawny owl is a woodland species, the short, rounded wings an adaptation to hunting within the closed woodland cover. They do, however, sometimes occupy other habitats, including urban parks and larger suburban gardens, so long as there is sufficient tree cover. Urban owls tend to take a greater proportion of small birds in their diet than woodland pairs, which feed predominantly on mice and voles. Tawny owls can be surprisingly adaptable when it comes to food, supplementing their diet with earthworms, which may be taken from the surface of garden lawns on damp nights, and larger insects. Bizarrely, they have also been recorded taking frogs and fish, the latter including an astonishing record of a bird that started to raid a garden pond to feed upon the goldfish contained within. This does underline that the tawny owl is happy foraging on the ground and that it is not afraid of water. I do wonder if the local birds ever visit our lawn and pond, but perhaps the garden is too small and too enclosed for a visiting owl to feel secure.

The young owl has moved on within a few days and our sleep is no longer broken by the harsh notes hurled at us from the tall beech that stands just a couple of gardens away. It is wonderful to have such a bird visit us, but we are both glad that it has left to find a territory somewhere else. We'll hear tawny owls again later in the year, but these will be the resident pair farther down the hill, their calls softened by the distance and all the more pleasing for it. Another bird whose nocturnal calling we often hear, especially through the summer months, is the oystercatcher. Although we are a good half an hour from the coast, we have several local pairs that breed on the tilled land that surrounds the town. Some of these birds overfly the garden, something I've mentioned in passing before, their piping call echoing across the still night air. Now that we are moving towards autumn, there will be other birds passing overhead at night, including passage waders making their autumn migration south from breeding grounds located far to the north. There is the possibility of passing whimbrel, godwits and sandpipers, all adding to the nocturnal soundscape and reminding us that there is a lot going on above and around our urban existence.

One Saturday morning, while working in the study with the window open, I hear the contact call of a chiffchaff coming from the garden. Scanning from the window with my binoculars, I finally spot the bird, working its way slowly through the vegetation that still hangs thick on the trellis though now well on the turn. The bird is looking for insects, a good number of which are still active in the late-season warmth. Like the waders this is likely to be a bird on passage, though its origins are perhaps less certain. Many pairs nest locally along the old railway line and on the common, and there is a chance that this individual has come from there. The chiffchaff is not alone and I see that it is loosely attached to a flock of tits, mostly blue and long-tailed, that are passing through the garden in an extended formation. Such flocks are typical of this time of year, the breeding season over and the restrictions that this brings now relaxed. These small birds can come together in mobile parties that wander in search of feeding opportunities. Being in a flock not only increases your chances of finding food, it also reduces your individual chances of being predated – the old safety in numbers argument.

When you watch a mixed species flock such as this, it quickly becomes apparent that the different species forage in slightly different places, thereby reducing the competition between them. The larger great tits spend more time on the ground, while the blue tits work the branches; the long-tailed tits and goldcrests are more agile, being smaller, and so work the finer branches and twigs. With its stout bill, the great tit is more likely to take seeds, while the goldcrests and long-tailed tits largely stick to small insects and spiders. These are also favoured by the chiffchaff. Such flocks are also worth scanning in case they hold any more unusual visitors. There is always a chance of a rare yellow-browed warbler, a species related to the chiffchaff but more strongly marked and with a different call. This used to be a very rare visitor to the UK but it has become much more common now, and good numbers are recorded most autumns. Other possible autumn rarities include pied flycatcher, wryneck and black redstart, all of which have turned up in UK gardens during autumn passage.

These mobile tit flocks, which are such a feature of autumn and winter, add to the sense of change. No longer is the pattern of visiting birds limited by the distribution of their breeding territories; now it is shaped entirely by weather conditions and the availability of food. Birds that have been breeding in the local woodlands or farmland hedgerows may increase their use of the local gardens, particularly once the natural stocks of seeds and berries are depleted. Indeed, increasing numbers of individuals pass through the garden now, stopping to take seed from the garden feeding station or to seek out insects in the shrubs and bushes.

* * *

A run of wet and windy days accompanies the passing of September's mid-point, the remnants of huge storms that hit America's east coast and which have now tracked across the Atlantic. Despite having lost most of their power, the winds still buffet the house and thrash leaves from the trees. I am reminded of the Ted Hughes poem 'Wind', which so beautifully captures the sense of overwhelming power contained within such storms. One morning, in the aftermath of such a storm, I watch a flight of lesser black-backed gulls, tracking west high above the house. The birds make slow progress as they fight their way forwards into the wind. Birdwatching friends have their eyes to the west too, aware that such storms can deliver American birds to our shores. Vagrant waders, warblers and thrushes, caught by the storm somewhere on the east coast of America – along which so many of these birds migrate south – are transported across the Atlantic to make landfall in Ireland, the north of Scotland or in the southwest of England. For many years, it was the Isles of Scilly, just off the tip of Cornwall, that were best known for these American vagrants, attracting an annual influx of birdwatchers hopeful of finding their own 'Yankee' rarity. Now, with a change in the pattern of these storms, we see the islands to the north of Scotland scooping an increasing share of these arrivals. Although I have made a single autumn trip to the Isles of Scilly, I prefer to restrict my birdwatching to much closer to home. There is reassurance to be found in the local seasonal pattern evident in the changing bird communities, as summer becomes autumn and then transitions into winter. That local focus, derided by some as parochial, is comforting and gives me more pleasure than that which might be gained from dashing off to see a bird that is far from where it should be.

Work in the garden shifts to cutting back and clearing, renewing the vegetable beds and preparing for the coming winter. My activities attract the attention of one of the local robins, whose sharp eyes remain alert to the soil-dwelling invertebrates that are revealed as I turn over the soil and lift the remnants of spent crops.

The recent rain means that the soil is soft and easy to work with the fork; it also means that the earthworms are close to the surface, something that soon attracts a female blackbird. She is much more wary than the robin, keeping her distance from me and choosing to feed at the far edge of the vegetable bed. Like me, she is turning over the soil, flicking away leaves as she searches for worms and grubs. Settled into my routine, and focused on the task in hand, I do not immediately notice the blue tits that are busy exploring the fence panels that surround this section of the garden. One of the tits hovers briefly at the corner of one of the panels, before grabbing hold of the wood with its strong feet to gain a steady purchase. The bird then probes a small gap, formed where the panel's upright joins the top section, and I can hear the noise of its beak as it tries to enlarge the aperture; it is after a spider. Despite the bird's persistence it leaves without its prize, the spider safely beyond reach and unharmed, even though its web has been destroyed.

I often see blue tits foraging for spiders in this manner, particularly early in the year when they turn their attention to those whose webs adorn our old window frames. I am always impressed by the way in which the birds are able to gain a foothold on the brickwork around the window, before they lean in to peek at the mat of web that protects a spider and her cluster of eggs. The blue tit's powerful grip is well known to bird ringers, and often surprises young trainees holding one for the first time. It is incredibly strong, and something that also enables these small birds to feed on wire mesh feeders and to cling to the underside of a coconut shell packed with suet and other treats. The strength in their grip reflects the blue tit's more natural feeding sites within the branches of trees and shrubs, and it is really only matched among similar-sized birds by great tit and nuthatch. The other thing that surprises trainee ringers is the power of a blue tit's beak. Despite its small size, the beak can pinch with significant force, much to the discomfort of the hand in which the bird is being held.

While there are birds using the garden, foraging in the hedgerow and shrubs or visiting the pond, the feeding station itself remains little used. There is so much natural food around that the birds do not need to turn to the seeds on offer in the hanging feeders or on the bird table. Even the woodpigeons seem to have found feeding opportunities elsewhere. This serves to underline that the food that we provide is used by the birds on their terms, rather than ours. It will be a few more weeks until we see an increase in feeder use, as the autumn 'trough' ends and the focus shifts back to the supplementary food on offer.

* * *

An email from a friend in Scotland brings with it a photograph of a bird new to their garden: a nuthatch clinging to the peanut feeder, with its rich panzer-grey plumage, striking eye-stripe and robust dagger-like bill. Although the nuthatch is very familiar to me, and recorded from each of the gardens that I have owned over the years, this is the first time that my friend has had one in their garden. It is also the first one that they have seen in Scotland: little wonder that they are so excited. The nuthatch has a southerly distribution within Britain, the species most abundant in the wooded landscapes of southern England and south Wales. It is largely absent from the open landscapes of the Fens and southeast Yorkshire, as well as from upland habitats. Over recent decades, however, the nuthatch's range has been expanding northwards, with breeding first recorded in the Scottish Borders back in 1989. Since then, the colonisation of Scotland has continued, the nuthatch population doubling here every two to three years, and the species is now established around Stirling in Central Scotland, with a scatter of breeding records to the north of this. BTO Garden BirdWatch records show the garden report rate for this species in Scotland increased from virtually zero in 1995 to one in ten gardens by 2013. It is likely that our increasingly mild winter weather is contributing to the movement north, but the availability of food at garden feeding stations may also be playing a role in what we are seeing.

Although broadleaf deciduous woodland is the favoured habitat, with oak (both common oak and sessile oak), beech and sweet chestnut the favoured trees, nuthatches will also occupy coniferous woodland – though at lower densities – and parkland, assuming the trees are not too widely spaced. Nuthatches are rather sedentary in habit, and so the range expansion that we have seen here in Britain is likely to have been driven by the movements of young birds during the period of autumn dispersal. The movements made by young nuthatches during August and September are thought to represent a second wave of dispersal, the individuals involved often making much longer movements than seen earlier in the season, when chicks first move away from the nest. Although not evident here in the UK, the autumn dispersal movements seen elsewhere in Europe sometimes see very large numbers of individuals moving significant distances. Such movements may be driven by food supplies, in the form of tree seeds, which, as we have seen elsewhere in this book, can be much less abundant in some years than in others. Interestingly, nuthatch is one of the species in which we see an increase in the use of garden feeding stations in those years when the crop of beechmast is low.

Another factor in the distribution of nuthatch populations is the fragmentation of favoured woodland habitats, something that is known to be a problem for a number of woodland bird species. As larger areas of woodland have become more fragmented, each block remaining has become smaller and more isolated from its neighbours. Isolated woodlands support fewer nuthatch pairs, with more suitable breeding territories remaining unoccupied. Plans for increasing woodland area in the UK, with many new woodlands now being planted, are likely to increase the availability of suitable habitats for this striking species over the longer term. However, given that it is mature deciduous woodland that is favoured, it is going to be some time before we see the benefits of the woodlands being planted now.

Landscape change is a particularly important driver for many bird populations, and we have already seen how it has altered the fortunes of 'farmland' birds like yellowhammer and tree sparrow, and 'woodland' birds like siskin, lesser redpoll, and now nuthatch. The increasing levels of urbanisation, as well as the associated demands on land for food and recreation that a growing population brings with it, have seen further change in our bird populations. As we have seen, it is those species that tend to be more opportunistic in nature – as opposed to being specialist in their habitat requirements – that have been best able to adapt to the built environment. While the nuthatch is not a species that you would associate with urban landscapes, it has almost certainly benefited from the increasing numbers of people providing food in their suburban and rural gardens, especially where these are located close to those areas of deciduous woodland used by breeding nuthatches. Unlike woodland-breeding tits, nuthatches have not taken to nest boxes to the same degree and they remain a relatively rare user of this resource.

* * *

Despite the lack of activity at the hanging feeders, there are still a few birds visiting each day to take sunflower seeds. One morning, late in the month, a chaffinch is perched on the feeder nearest to the kitchen when a sparrowhawk arrives. The larger bird skims low over the fence, the arc of its trajectory seemingly taking it towards the bird table. However, with the merest hint of movement the sparrowhawk changes direction, its focus now on the chaffinch that has already left the feeder in a blur of wings. In panic, the chaffinch makes a wrong decision and flies into the kitchen window with an audible and sickening bang. The sparrowhawk swoops past the window, narrowly missing a similar fate, and disappears around the corner of the house. Fearing its return and feeling complicit in what has happened to the chaffinch, I rush outside. The bird is on the ground beneath the window, sitting upright but clearly stunned. I pick it up, its tiny warmth cradled in my cupped hands, and find a small box into which it can be placed in the hope that it recovers. I leave the box on the side in the utility room, where there is warmth and the bird can remain undisturbed by the other members of the household. Although its eyes were half-closed, there was no other external sign of injury, but it is what has happened inside this tiny scrap of life that will determine whether or not it survives the next few hours.

Over the years, I have encountered a number of window-strike victims, some of which are killed outright by the collision, others which make a recovery. More often I see the impressions left on the glass by a bird that has collided with the window and then either flown away or been removed by a predator. The impressions left by pigeons are the most obvious because of the almost greasy nature of their plumage. Viewed against a dark background, the print left captures a detailed impression of the bird involved, each wing feather clearly visible and captured at the moment of impact. Sometimes you can make out where an eye or the bill is, its presence a negative space within the print, only visible because of the feather detail around it. Once, when I was a child, we found the perfect imprint of a tawny owl captured on our glass French doors. You could clearly make out the facial disc and the two large, forward-facing eyes, each a void within the feathering of the disc itself. On another occasion, the bang of a male bullfinch hitting our kitchen window was followed a split second later by a louder and more shocking bang, as a female sparrowhawk hit the glass behind her. Both birds were killed instantly, their still warm bodies a stark reminder of the fragility of life and of our impact on the creatures with which we shared the garden.

The chaffinch, a young bird, is evidently recovering well. As I open the door to the utility room, I can hear it scrabbling about inside the box. Carefully, I slide my hand into the box and secure the bird in 'ringer's grip' – a way of holding small birds that avoids the risk of damage and seems to placate them because of the small amount of pressure felt over their closed wings. Both eyes are now fully open and the bird is alert to what is going on around it. Confident that it needs no further attention, I take the young bird out into the garden, away from the house, and release it close to the hedge. It flies up into the nearest cover and I leave it to orientate itself in the hope that it will fly away none the worse for its experience. Being flushed from a feeder by a predator does appear to be one of the contributing factors behind collision with a window, and it is for this reason that feeders should not be placed too close to glass.

I see a sparrowhawk again the following day and assume that this is the same bird, our garden part of its regular beat as it searches for an unwary finch, tit or thrush. I also see chaffinches at the feeders and bird tables, their numbers increasing a little in response to a change in the weather and drop in temperature. In with the chaffinches are a couple of greenfinches, both adult males judging by the strength of colour in their bright yellow wing-markings. These run along the length of the central shaft, starting near the feather's base, where the yellow extends from the shaft to the feather's outer margin, but narrowing in width towards its tip, such that the last centimetre of the feather lacks the yellow altogether. On the folded wing, with each feather overlapping that of its neighbour, these narrow lines of yellow come together to form a more extensive panel. The lines of yellow are narrower in the female, and the yellow less intense, but in an adult male they are as bright and as bold as the yellow in any child's paintbox, enhancing the mossy-green tones of the wider plumage and contrasting with the darker feather tips. It is the presence of the hanging feeders that brings these birds close enough to view in this level of detail, enabling me to take in the differences in plumage that might separate male from female or adult from youngster. It is one of the reasons that I put out food for wild birds.

Not everyone who puts out food for the birds does so for similar reasons. Many people provide the food in the hope that it is helping their avian visitors but without showing a wider interest in understanding which birds visit and why. Others are fascinated by the birds that visit and keep detailed notes or participate in citizen science projects in order to find out more. Such differences need to be viewed against the backdrop of wider societal differences in who puts out food for birds. It is known, for example, that the practice of feeding wild birds tends to be linked to household income. In general, participation in feeding garden birds and the numbers of bird feeders used both increase as household income increases, when viewed across urban communities. Research carried out in North America has revealed that older people are more likely to put out food for birds than young people, that women are more likely to feed than men, and that those providing food tend to have achieved higher educational qualifications than those who do not feed.

In addition to such patterns, there are others that relate to cultural practices – feeding is more common in Britain, Germany and the Netherlands than it is in France or Spain. There is also evidence that food provision is shaped by geography, such that the provision of food is greater in areas where the winter weather conditions are harsh, compared to areas where they are relatively mild. There is also a difference between urban and rural communities, though it is worth noting that many of the factors already mentioned also vary between urban and rural communities. The foods provided may also differ, something evident in a study carried out in Poland where researchers found that seed and waste food were provided significantly more often in urban communities, while products based on animal fats were more common in rural communities. Over wider geographic scales, we see similar differences. In southern North America, for example, 'nectar' feeders using sugar solution are commonly used to feed visiting hummingbirds, while in Australia, it is high-protein foods – like cheese and meat – that are more commonly provided.

It would be fair to say that we do not have a good understanding of the nutritional requirements of garden birds, at least not in the sense of understanding what the foods that we provide at bird tables and in our hanging feeders actually provide. We assume that the seeds that we are providing, such as sunflower hearts, millet and grains, provide similar nutrients to those that the birds would get from feeding on native plants, shrubs and trees. The provision of suet-based products, such as fat balls and energy bars, while favoured by many garden bird species, are clearly different from the invertebrates and berries that other garden birds take. It has only really been over the last few years that researchers have begun to examine the potential impacts of feeding garden birds, turning their attention to the nutrients present in particular foods and seeking to understand whether their provision has consequences for individuals. There is still a very long way to go with this work, and it is hampered by the difficulties in studying such effects within free-ranging wild populations, but it does serve as an important reminder that providing food may not always be beneficial. Having said that, it is worth remembering that our activities more broadly have greatly changed our countryside and impacted on the creatures with which we share it. Some of these activities may have proved beneficial for some species, but for others they may be detrimental.

This then raises the question of whether or not feeding is a good thing. On balance, it appears that the benefits significantly outweigh the costs, but we don't have all the answers yet and it remains difficult to separate out the effects of feeding or not feeding from other factors, such as land-use change or climate change. We do know that the provision of supplementary food can increase overwinter survival rates, enabling more individuals to make it through to the following breeding season, and that feeders provide a reliable food source for young birds as they gain independence. While food provision can make a difference at the level of the individual bird – perhaps determining whether it survives a particularly challenging winter night or not – it is much harder to determine the role that it plays in populations or within communities comprised of many different species. There are also questions around its impact at a wider scale, at the level of a food web for example. We know from research carried out here in the UK that populations of certain invertebrate species, such as aphids and beetles, may be reduced in those gardens where bird feeders are positioned. Presumably the bird feeders attract more birds, which then increases the predation pressure on the aphid and beetle populations present. In the case of the aphids, this is good news if you grow roses or broad beans, but not good news if you are an aphid.

It is when I am faced with the complexities of these relationships around feeding wild birds that I tend to focus back onto the individual act, my provision of food for individual birds who spend all or part of their year in my garden. This is brought home to me at those moments when I see the young blackbirds from the nest near the greenhouse visit the garden feeders for the first time, or that young chaffinch that collided with the window earlier in the month. Each of these birds is an individual, like me, making its way in the world and facing the challenges of simply being alive. Each is a potential predator and prey, at risk of becoming the victim to some disease or misfortune, but equally with the potential to create new life and continue a line of blackbirds or chaffinches, a line that began long before we started providing food and which will end who knows where at some point in the future. Right now, here in my garden, we are interacting through the food that I provide, through the way in which I manage my garden and because I am living here at this point in time. When viewed like that, the relationship feels a whole lot more personal and immediate, freed from those wider concerns. Over the coming weeks, as autumn shifts into winter, the garden community will change and other birds will be drawn in, the food that I provide having a hopefully positive influence on a wider circle of individual lives.

October

It is during October that activity at the garden feeding station increases again after the late-summer, early-autumn lull. This seasonal lull is perhaps most obvious in the weekly data collected by BTO Garden Bird-Watchers, which results in a series of weekly 'reporting rates' for more than forty familiar garden bird species. These data show that the reporting rate for blackbird drops from in excess of 90 per cent in July to a low of 74 per cent during September, before increasing in October to arrive back at 90 per cent by mid-November. What is striking about this seasonal trough is its regularity, year after year, the timing shifting only very slightly between years. Part of the reason for this seasonal pattern is the changing availability of natural foods. The abundance of late summer and early autumn is coming to its end, the seed and early berry crops depleted by the birds, insects and small mammals that have been making the most of them over recent weeks. Another factor that comes into play during October is the arrival of large numbers of thrushes and finches from overseas. Many of these arrivals, in fact most of them, will be of species for which the UK already holds significant breeding populations, for example greenfinch, chaffinch and blackbird.

October is traditionally the month when the autumn migration of birds brings unusual visitors to our shores, perhaps individuals that have become disorientated by the weather conditions, drifted west across the North Sea, or for some other reason headed in the wrong direction when they set off on migration. While the focus of most birdwatchers will be on these unusual visitors, some of which do end up in gardens, there will be very many more thousands of common birds making their regular migration to wintering grounds here in the UK. Most will go unnoticed, joining others of their kind that belong to our resident breeding populations, but they can be visible when they first arrive, especially if the weather conditions conspire to drop large numbers of individuals on the coast. Winds from the east and clear skies over Norway will see the evening departure of many small migrants from the Norwegian coastline. If the eastern UK is covered with night-time cloud and rain, then an early morning trip to the coast can reveal bushes full of thrushes and goldcrests, grounded by the weather and in need of food to fuel the next leg of their journey. It is quite a sight to encounter these birds, newly arrived and packed into the sparse cover that is typical of many stretches of the East Anglian coastline. There is something particularly intimate in being able to sit in the dunes just a few feet from feeding goldcrests: tiny scraps of life that have crossed the North Sea and faced the challenges of both the distance and the elements.

Not all of those that set off make the crossing successfully. In some years, dozens of corpses may be found along the tideline: individuals that ditched in the sea exhausted or were overcome by particularly challenging weather conditions. This autumn, the heavy overnight rain appears to have caught out large numbers of redwings and other thrushes, the social media posts of those who walk the coast just a few miles north of here documenting their sad fate. One evening a few days later, while walking up through town and back towards the house, I hear the soft 'see-ip' call of a redwing passing overhead in the clear and starlit sky. More calls follow, suggesting a flock on the move in the darkness above, each individual keeping in touch with the others through these thin, far-carrying notes of connection. These are my first redwings of the winter, an arrival of birds from Norway or beyond heralding the changing season and emphasising the scale of connection that exists between here and elsewhere. It is amazing to think that the skies above this small Norfolk town link together these wintering thrushes with those summer-visiting swifts that will by now be somewhere in the heart of Africa, a world away from here.

Many small birds migrate at night rather than during the day, the night-time conditions more energetically efficient (the birds are less prone to overheating) and the risk of predation reduced, though not removed entirely. Migrating at night, small birds may be caught by hunting peregrines as they pass over our towns and cities. Farther south on their journey, as they pass through Spain, they may be taken by greater noctule bats, a species that has recently been found to specialise on small migrating birds. Given the vast numbers of small birds moving south during the autumn, it is perhaps not surprising that at least one nocturnal predator specialises in their capture. Migrating at night also enables birds to use the position of the stars to aid their navigation, one of a number of navigational aids available to migrating birds. Others include visual landmarks, the Earth's magnetic field and the position of the sun. There is also a strong genetic component to avian migration, such that young birds inherit a particular set of behaviours from their parents, something ably demonstrated by work on blackcaps and other songbird species.

It is not that late in the evening, so I wonder where these redwings set off from. Are they birds that have been feeding up locally during the day and which are now beginning the next leg of their journey, or have they already covered a substantial distance? One redwing, ringed on Merseyside during its autumn migration, was found freshly dead in Spain just four days later, having covered at least 1,136 kilometres in that time. If it only moved at night, then it would have been averaging just shy of 300 kilometres each night. This is exceptional, however, and a more typical movement for a small nocturnal migrant is thought to be around roughly half this distance, at 170 kilometres per night. The speed of avian migration is usually higher in nocturnal migrants than diurnal ones, in long-distance migrants than short-distance migrants, and for species that begin their migration earlier in the autumn than those that set off later. Of course, other factors come into play, such as weather conditions, the location of stopover sites (where a bird can refuel), where on its journey a bird is, and the size of the fat reserves that it is carrying.

Migration is a complex business, with a variation in approach and known to differ between species, between populations and between individuals. Out there, in the darkness of this autumn evening, hundreds of thousands of birds are moving, some in flocks, some on their own. For the most part it is an unseen spectacle, one of nature's marvels and something about which we are still learning. What we do know comes from the records of birdwatchers, collected together in annual reports published at the county level, and the efforts of those studying migration through the tools of bird ringing and tracking devices. Our understanding has come on rapidly thanks to the introduction of tracking devices, such as the tiny geolocators fitted to black-caps, nightingales and other small birds. Such small devices cannot transmit their location, but instead hold this within the device. Only if and when the bird is recaptured can the data be downloaded and the bird's movements revealed. Larger migrants, roughly from the size of a cuckoo up, can be fitted with devices that report the bird's position at intervals throughout its journey. Not only does this mean that the birds don't have to be recaptured to secure the information collected, but also that a more complete picture of the movements is gained. With geolocators, you only get information back from those birds that have been successful in their migration, but for satellite tags you also get information from those birds that die during migration.

Such knowledge is of great value because it shows where, when and why birds are dying on migration. Knowing this can reveal the factors that may be driving the decline of a species, something nicely illustrated by work on UK cuckoos carried out by Dr Chris Hewson of the BTO. Before the use of satellite tracking to follow the migration of UK cuckoos, the small amount of information collected through ringing had suggested that UK cuckoos migrated southeast, passing around the eastern end of the Mediterranean and probably wintering somewhere in the eastern half of Africa.

As soon as data started to come in from the first cohort of UK cuckoos to be fitted with satellite tags, it became clear that the birds used two different routes to reach the same wintering grounds, located in Central Africa around the Congo Basin. Some of the birds did travel southeast, but others took a westerly route that saw them pass down through France and Spain, and skirt West Africa before turning east towards Congo. Even more surprisingly, all of the birds returned by the western route in spring, regardless of the route used the previous autumn. As successive cohorts of cuckoos were tagged, it was revealed that the choice of route had implications for whether or not a bird was likely to make it to the wintering grounds successfully. Those birds that took the western route suffered from higher levels of mortality during their autumn migration, the birds seemingly struggling to refuel effectively in southern Spain. This may have been linked to changing land use in the region, together with extended periods of drought during the autumn migration period. Interestingly, cuckoos breeding in different parts of the UK tended to show differences in the route used; those birds from the north and west of the UK (where cuckoo populations are stable or increasing slightly) tended to use the eastern route, while those breeding in the south and east (where the populations are in sharp decline) used the western route. Could the choice of migration route be behind the regional pattern of population change witnessed in this species? It seems likely. Of course, this is only part of the story, because the weight of the tags means that they can only be fitted to male cuckoos – the larger of the two sexes. Females (and indeed also young birds) might do something different, so we will need to wait until smaller tags become available before we can develop a more complete picture of the link between migration behaviour and population decline in this species.

* * *

Evidence of birds on the move can be seen in the flock of siskins that puts in an appearance during a lunchtime walk along the river, not far from the house. I hear the flock before I see it, an overlapping chorus of twittering calls that underlines the very vocal nature of these feeding flocks. The flock moves towards me, cascading down into the upper branches of some riverside alders, heavy with their characteristic cones. The shallow valley of the upper Waveney is dominated by wet ground, and this section of the river is lined by wet woodland, a good deal of which is alder. There will be sufficient seed in these cones to keep the siskins here for some weeks, assuming that they do not move on elsewhere, and it will not be until early in the New Year that they are likely to put in an appearance at my garden feeding station. I scan the flock and strain my ears for the chance that a few lesser redpolls might be present alongside the siskins, but no luck.

Also evident locally are the goldfinches, some of which have been frequenting the sunflower hearts in my hanging feeders. Like the siskins, the presence of the goldfinches is often revealed by their call, a series of sharp notes that tinkle like shards of glass. Again, there is something of a conversational quality to the calls of a feeding flock, or 'charm' as it is sometimes known. From late summer, goldfinches may be seen feeding on thistles and other seeds, preferring those that are only half-ripe and still in their milky stage. Finding seeds that are at the right stage may be one reason why goldfinches are so mobile, their small feeding flocks continually on the move from one patch of suitable vegetation to another. To watch them feeding on the seed-heads of thistle or burdock is a delight, their agility unmatched by our other common finches. The agility of these birds is lost when they visit my hanging feeders, the feeding perches a far less challenging proposition than the delicate stems and seed-heads of scabious, hawkbit or hawkweed. When the large, oil-rich seeds on offer in my hanging feeders are viewed against the much smaller and seemingly much less profitable wildflower seeds available within the wider countryside, one wonders why these birds don't make even more of garden feeding stations. There is a strong seasonal pattern to garden use, and goldfinches feature in many gardens throughout the year, but clearly much of the population only turns to supplementary foods during the winter months, when natural seed supplies are at their lowest. It is during the winter months, particularly during the first quarter of the year, that goldfinches also make use of tree seeds, favouring alder, birch and pine. During the summer months, seeds of various members of the daisy family – the Compositae – dominate the diet, while in autumn it is those of thistles. Right now, as some of the birds that I have been watching over recent days neatly demonstrate, teasel and burdock come to the fore.

Another pattern evident in these goldfinch flocks is the change in the number of birds present within a flock as the winter progresses. Autumn flocks tend to be larger than those seen later in the winter, although this is very much dependent on the amount of food available. At some sites, including some garden feeding stations, sizeable flocks can gather together to feed. This pattern is also seen in some of our other common finch species, underlining the importance of food availability in driving movements and feeding behaviour. One difference, certainly evident within the wider countryside but absent at garden feeding stations, is the tendency for goldfinches to feed in single-species flocks, rather than join mixed flocks of other finch species. This reflects their more specialised diet, something that tends to keep them away from farmland strips of 'game cover', where linnets, greenfinches, chaffinches and bramblings may come together to feed. It may be for this reason that I tend to think of goldfinch as being a little different from its relatives: a more refined finch, at least away from the garden feeding station where it can be a bit of a bully.

The garden feeding station itself has become the focus of a young grey squirrel, which has discovered that there are some easy pickings to be had. For some reason, the young animal has been favouring the sunflower heart feeder closest to the house, and some distance from any trees or other cover. If it happens to be on the feeder when I walk out of the back door, the squirrel leaps onto the fence and then climbs up the brickwork, only pausing when it reaches the guttering above the spare room. Feeling secure in its lofty perch, it churrs at me in alarm, no doubt frustrated that it has not been able to feed in peace. Although the key parts of the feeder are made from metal and not plastic, I do wonder if the animal might be able to break its way into the tubular design. As it stands it is only able to extract seeds individually, and with some effort, from the feeder ports, designed to deliver seed to more delicate visitors.

Over the years, I have tried several different approaches in order to deal with troublesome squirrels, from more robust feeders, to the use of poles, baffles and deterrents. I have learned that grey squirrels are resourceful, and able to cover a surprising distance when jumping from a standing start. A feeder on the top of a pole, protected from below by a baffle, only remains out of reach to a squirrel if the pole is placed a good distance away from any other structure, such as a tree or fence, from which the squirrel can launch itself at the feeder. Similarly, hanging a feeder from a branch, with a baffle placed above the feeder, only works if the feeder is some distance from the trunk and there are no nearby branches from which the squirrel can launch a sortie. The use of a chilli-based powder appears to act as a deterrent – like us, grey squirrels have taste receptors sensitive to chilli, which are absent in birds – but it isn't something I have ever felt comfortable using. Another alternative, which does seem to work, is to provide the squirrels with their own feeder, full of peanuts and resembling a bird box but with a Perspex front and a lid that can be lifted up. Sometimes, these can be so successful that you find yourself greatly increasing the quantity of peanuts that you order each month. Still, it can be fun to watch a squirrel climb inside the feeder, only its tail left hanging out while it stuffs itself with the peanuts provided. A final option is a spring-loaded feeder: these either use the weight of the squirrel as it clings to a feeding perch to close off access to the food within or to twist the feeder sideways at speed, which results in the animal being flung from the feeder and onto the ground below.

The grey squirrel isn't the only mammal to visit the feeding station this month, as a camera trap reveals. Set watching the ground beneath the feeders, the camera picks up the nocturnal visits of wood mice and bank voles. That these are wood mice and not yellow-necked mice cannot be determined for certain from the footage obtained, but we are sufficiently far north within Norfolk to be beyond the current range of yellow-necked mice here in East Anglia, though that may change. The bank voles are particularly welcome, being one of my favourite small mammal species with their russet coat, blunt face and neat appearance. Field voles have a more untidy appearance because of the longer guard hairs in their coat. Both the wood mice and bank voles are feeding on fragments of seed that have been dropped by birds feeding on the feeders suspended above. I wonder just how far they have come and ponder where in the garden they are breeding. Although regular inhabitants of many rural and suburban gardens, it is known that they are unable to maintain populations in more built-up areas because of the higher densities of cats found in these urbanised landscapes.

Less welcome is the common rat that puts in an appearance, not because I have anything against the species but because their presence can alarm neighbours and lead to requests from local councils for householders to cease putting out food for wild birds. The risk of attracting large numbers of rats can be reduced by keeping feeding stations clean, removing spilt food and making sure no food is left out overnight. Keeping the garden tidy of rubbish can also help, reducing the cover under which a colony can become established. The presence of chickens can increase the attractiveness of a garden to rats, so again good hygiene practices and ensuring that the chicken run is inaccessible to rats are wise precautions. The rat in question only appears on a couple of nights so it must either have been caught by one of the local cats or been passing through. It is something to watch for though, especially as I am not the only one in this group of households to feed wild birds, and I make a mental note to make a proper small mammal feeding station, sited within a wire mesh cage that will only allow access to the smallest of small mammals.

* * *

One afternoon, pottering in and around the greenhouse, I pause to watch a blackbird that has been bathing in the shallows of the pond and is now busy preening. The bird is working through its different feather tracts with its bill, occasionally twisting its head back towards the base of its tail, where the preen gland is located. This gland, known by academics as the uropygial gland, produces oily secretions that are used to maintain feather condition. The secretions have been shown to have antimicrobial properties that inhibit the growth of feather-degrading bacteria, together with insecticidal properties that help to control the chewing lice that can damage feathers. The size and shape of the gland varies between the different garden bird species, and there is also a degree of variation evident between individuals of the same species. For example, house sparrows with larger preen glands show less feather wear than those individuals with smaller glands, underlining the role that its secretions play in feather maintenance.

I sometimes see birds visiting the garden that show some kind of plumage abnormality. The abnormalities often take the form of missing feathers, the bird having a bare patch on its head or elsewhere on the body. While the loss of feathers is part of the moulting process, which itself can sometimes go awry, it can also be the result of disease or parasites. In some cases, the loss of feathers is the result of a fungal infection, the fungal threads infiltrating the skin and the feather follicles, hindering growth. Although I have not encountered any severe cases of feather loss in the birds using my garden, I have been sent photographs of individuals showing alarming conditions: these included a robin that had no feathers anywhere on its head or neck, but the rest of its plumage was untouched. In other respects, the bird seemed healthy and the observer reported that it had continued to visit their garden in this condition over many weeks, feeding normally and chasing away other robins.

While the loss of large numbers of feathers is uncommon, you sometimes see birds with small amounts of damage to individual feathers. Such damage is usually the result of chewing lice, which feed on the feathers themselves, though some species of lice instead obtain blood from growing feathers or live within the quill and feed on its pith. Other damage may be the result of a period of nutritional stress, something seen fairly commonly in the feathers of young birds, newly fledged from the nest. If these young birds experience a period of food shortage as a nestling, while their wing and tail feathers are actively developing, then this can produce what is known as a fault bar. The bar, which runs across the entire feather tract (so across one or both wings, and/or the tail), appears as one or more horizontal lines. In the worst cases, the feathers can be so weakened at this point that they break under the strain of flight. The condition may sometimes be seen in adult birds, occurring where the individual experienced a period of nutritional stress while undergoing moult and replacing its wing or tail feathers.

The blackbird appears to be in excellent condition: an adult male, its soft black plumage is sleek and neatly aligned by the busy bill. Only the bird's tail looks less than perfect, perhaps because it has been submerged in the pond but maybe because it has got a bit battered moving through vegetation. Battered tails certainly seem to be a feature of blackbirds in late summer, especially the females who have spent a lot of time incubating in the cramped conditions of the nest. Having raised three or even four broods of chicks, it is little surprise that an adult blackbird can look a bit scruffy. Once the demands of raising a succession of broods are out of the way, the adult can secure sufficient resources to carry out its annual post-breeding moult.

Moult is an energetically expensive activity, so its timing needs to fit in with an individual's other responsibilities. While this is rather straightforward to achieve for a resident bird like a blackbird, whose year is divided up between the breeding season, the autumn moult and the challenges of winter, for those birds that migrate the timing and pattern of moult become more complex. There can be an urgency for these migrant birds to return to their breeding grounds, secure a mate and breed, before leaving again. Some manage to squeeze in their annual moult between the end of the breeding season and the beginning of the autumn migration. Others wait until they have arrived on their wintering grounds, while some start their moult before they depart, suspending it for the period over which they are moving south, and then restart once they arrive on their wintering sites. The decision as to which strategy to adopt is shaped by how much time they have before they depart, how far they have to fly and a number of other factors, including age and taxonomy.

The process of moulting and replacing feathers is something largely hidden from casual observers, in part because it is hard to see when a bird has its wings or tail folded closed. Having said that, if you watch a large bird in flight above you, such as a crow, woodpigeon or bird of prey, you can often see the gap where a feather is either missing or actively being replaced. Those fortunate enough to see birds in the hand, perhaps through their participation in the national ringing scheme, get a much better appreciation of moult, its timing and how it progresses. Bird ringers record the pattern of wing moult for many of the birds that they handle, counting the number of feathers and then scoring the moult progression for each feather in turn. It is often possible to determine whether an individual feather has been moulted or not because of subtle differences in shape or background colour. In the tawny owl, a species that I handle fairly regularly, the pattern of barring on the wing differs between those of adult and juvenile feathers. Since tawny owls take several years to replace all of their wing feathers, and do so in a particular pattern, it is possible to age a bird by looking at the pattern of adult and juvenile feathers present. Another interesting element to tawny owl feathers is that the pigments that give them their colour have been found to fluoresce under ultraviolet light. The strength of the fluorescence declines as a feather ages because exposure to light breaks down the pigment; shining ultraviolet torchlight across the wing soon reveals the different ages of feather, making it easier to determine the pattern of feather moult.

It has been found that a number of familiar garden bird species have plumage that also shows considerable reflection of ultraviolet light. Given that most birds are visually sensitive to wavelengths in the near-ultraviolet spectrum, it seems likely that the two are linked and that individual blue tits, for example, will look very different to each other than they do to the human eye. Male blue tits are considerably brighter than females when viewed under ultraviolet light, especially on the crest. In fact, experiments demonstrate that female blue tits use the brightness of the male's crest when selecting a suitable mate. Females show a strong preference for males that have the brightest crests, with crest brightness seemingly indicating some aspect of male quality. One wonders what such features, only now becoming apparent to us, reveal about an individual blue tit to others of its kind. We know that plumage colour and brightness, and indeed the colour of a chick's gape, provide information on an individual's condition, suggesting that other individuals might use this information to inform the decisions that they make, such as who to select as a mate or which chick to feed.

Thinking about the blue tits and how differently they must see the world, reminds me that even though we have accumulated a great deal of knowledge about birds, their populations and their behaviour, there is still so much to learn. If we are to understand the impacts that we are having on their populations, then some of the knowledge yet to be gained is likely to prove important. The difficulty is in determining what we need to know if we are to make informed decisions about our activities and reduce or remove their impacts on the birds and other wildlife that live alongside us. Nature is complex, as the evolution of the signalling functions of plumage in blue tits proves. If something we do, such as the introduction of a new foodstuff to bird tables, changes the plumage characteristics of an individual, then what might the consequences of that be? One can imagine that the introduction of a foodstuff rich in antioxidants, which have been linked to plumage colour, might have consequences for an individual during its later life. If plumage colour or brightness is an honest signal of an individual's quality or fitness then that might be fine, but what happens if it isn't? What if we have somehow messed up the signals being given, such that they no longer provide an honest signal? Before we can address that concern, we need to understand how the signals work and identify the underlying factors that shape them. Put simply, why do some male blue tits have brighter crests and what is it about their genetic make-up and/or diet that gives them this brighter plumage?

* * *

It is during October that the change in day length really hits home, the shortening daylight hours curtailing the time spent in the garden. It feels like a time to hunker down and prepare for the coming winter, my own feelings mirroring the behaviour of many of those birds and small mammals that are busy stocking up on seeds and the remaining berries and fruits. A cache of sunflower seeds under some pots in the greenhouse suggests that a mouse or vole has been busy, the nearest bird feeder a few metres away. The seeds are mostly fragments, the discarded or spilt remains that collect beneath any feeding station. I wonder on the quantity of such remains that must fall to the ground each year beneath my feeders, then ponder on how important this 'manna from heaven' is for the mice and voles that use the garden. Elsewhere around the garden, as will be discussed in the following chapter, will be the caches of seeds stored by coal tits, jays and nuthatches. Some will remain undisturbed, lost and forgotten, while others will be found during the difficult months ahead.

A blue tit has been roosting in the porch, an open structure that affords sufficient protection from the elements for any small bird that huddles up on one of the cross-beams that supports the small, tiled roof. Two other pieces of timber lie parallel to one another, a few inches apart, running from the roof's apex to the cross-beam, and it is into the small space between these that the blue tit has secreted itself. For some reason, it seems to prefer the left-hand side of the door, even though there is an identical space on the opposite side. The bird's presence is revealed each morning by the addition of a few more droppings on the beam on which the bird perches. Each evening, when arriving home after dark, I check for the blue tit's presence and the reassurance that all is well in its world. On some nights it is absent, though after a couple of weeks it becomes clear that the bird must have other roosting options, the porch used perhaps every three days out of five. Even though it is still early evening when I arrive home, the blue tit is already roosting, head tucked in and the bird settled quietly. There's no using the porch light for now as it might disturb the roosting bird, waking it unnecessarily, or making it more vulnerable to a nocturnal predator.

I wonder if this roost site will still be suitable once we reach the colder nights of November or December. Perhaps it will prove to be too exposed, leaving the blue tit to find another option elsewhere. The energetics of roosting out in the open are challenging for a small bird like this. Being able to get out of the wind can make a big difference to the amount of heat lost and fat reserves used in an attempt to maintain body temperature at a safe level. Roosting birds like blue tit often employ a behaviour called rest-phase hypothermia, which involves a controlled reduction of body temperature in order to conserve energy during periods of high metabolic demand on the body's reserves. Being small, with a body weight of roughly 12 grams, blue tits frequently face low winter temperatures and reduced food availability; together these dual challenges make it difficult for the bird to maintain a positive energy balance through the winter months. The availability of a predictable food supply, in the form of the sunflower hearts provided in my hanging feeders, coupled with a suitable roosting site, certainly makes a positive difference.

Studies on great tits have revealed that individual birds actively explore a number of potential roost sites before settling on the one that they will use that night. This might be an open cavity, such as that being used by the blue tit above, or it might be that provided by a nest box, perhaps one that I cleaned out a few weeks ago, back in September. Of course, there are likely to be just as many natural cavities, such as those formed within living or dead timber. A study of such natural cavities, which explored their microclimate, found that the temperature within tree cavities was often several degrees above the ambient temperature recorded at the site. The size of the cavity's entrance hole was important in determining the temperature within the cavity: a small entrance hole being better than a larger one. Interestingly, cavities in healthy trees were, on average, warmer than similar-sized cavities in dead trees, though this difference was only evident at low temperatures and disappeared when the mean ambient temperature was higher.

Preferred sites offer both warmer and more stable conditions. Selection appears to be shaped by the weather, the bird selecting a site appropriate to the conditions it faces. This suggests that each bird will be aware of a number of different roost sites in the local area, and that there might be some degree of competition for the best of these. It might also indicate that there are benefits to be had by remaining in a particular area, gaining knowledge of the feeding and roosting opportunities on offer, so that this knowledge can then be used to help the bird survive the winter conditions. Again, as the month nears its end, I am reminded of the community with which I share this small plot, and how I manage the garden will create and hopefully maintain the roosting opportunities that these small birds will be using now and over the coming weeks.

November

I hear the sparrowhawk before I see it, the rush of the swooping bird as it drops onto its target the other side of our garden fence. Judging by the shrieking calls, it has caught a blackbird, something that is confirmed a few minutes later when the sparrowhawk takes to the wing and appears above the fence, powering her way towards the cemetery with the now dead blackbird in her talons. I can tell that this is a female sparrowhawk from her size. She is about 20 per cent larger than her mate and able to tackle prey up to half a kilo in weight, which would include woodpigeon and collared dove; the male's limit is about 120 grams, a little heavier than this blackbird. The difference in body size between the two sexes is seen in a number of other birds of prey, and indeed in some other groups of birds. In birds of prey, the degree of difference is more pronounced in those species that hunt fast-moving prey than those that don't. Known as reverse sexual size dimorphism, the female larger than the male, it is thought to be linked to hunting efficiency and the different roles of the two sexes.

A female sparrowhawk requires the energetic reserves to produce a clutch of eggs, and a larger bird can produce larger and more eggs than a smaller one. She can also defend the nest more readily and dominate her male partner more easily if she is the larger of the pair. The male's role is to hunt and to provision his mate and chicks with food. Hunting efficiency in birds of prey is increased where the predator is more closely matched in size (and hence agility) to its prey, something that would tend to favour smaller body size in male birds of prey that carry out the bulk of the hunting role during the all-important breeding season. There is one other benefit that emerges from this difference in size between the two sexes: it enables them to partition their prey base, reducing the degree of competition between the two, with the female taking larger species and the male taking smaller ones.

The spectacle of a sparrowhawk killing another bird on your lawn is a disturbing one, and something that feels very different from the sight of a blackbird killing a worm or a song thrush taking a snail. In all three instances, one creature is taking the life of another, but because we place greater value on the life of a bird than of a 'lower' organism we see it in a very different way. Nature is 'red in tooth and claw'. It can be brutal and uncompromising but we have tended to sanitise how we view the natural world, compartmentalising the different processes depending on how we value the organisms involved. The killing of a garden bird by a sparrowhawk can also feel more of an intrusion because the chances are that we have put out food for the house sparrow or robin that is the unfortunate victim. We see it as our fault and we view the sparrowhawk as the villain because it threatens the benevolent relationship that we have created with these visiting small birds. Some people go further than this, and blame the sparrowhawk for the wider decline in song bird populations, making a leap from witnessing an act of predation to some far greater impact. This is unfortunate, not least because it plays into the hands of those who want wider controls on bird of prey numbers because of their own commercial interests, namely the production of pheasants for shoots that attract sporting guns willing to pay large sums of money to show off their prowess.

It is certainly true that the populations of sparrowhawks and some other bird of prey species have been increasing, their numbers recovering from decades of persecution and the effects of agricultural chemicals, like aldrin and dieldrin, which saw the near collapse of food chains. As sparrowhawk populations have recovered, so they have recolonised former haunts and the species is now a common sight within our towns and cities. At the same time, we have seen the widespread decline in populations of many farmland birds, whose numbers have been falling since at least the 1960s. It is all too easy to suggest cause and effect, the increase in a predator's numbers and the decline in those of its prey, but it would be wrong to do this without doing the research and analyses necessary to determine that this is causation rather than correlation.

I have been fortunate to have been involved in some of this research, using long-term data from the BTO's Garden Bird Feeding Survey to look at the pattern of sparrowhawk recolonisation of gardens and its potential effects on the other garden bird species present. Our work, using complex statistical techniques, does find an effect but it is an incredibly weak one. This result supports many other studies, carried out using different data sets and looking at other bird communities, all failing to find any evidence to support the assertion that sparrowhawks are behind the long-term declines seen in species like yellowhammer, tree sparrow, starling and house sparrow. Instead, we need to look elsewhere, not least at the significant changes in land management and agricultural practices, at the growth in urbanisation and our growing impacts on the environment. Sparrowhawks and other birds of prey make an easy target, but it is our own actions that require attention.

The presence of a visiting sparrowhawk can shape the behaviour of other garden birds. Work carried out several decades ago revealed that adult great tits prefer to feed on hanging feeders that are located close to cover. Presumably, the nearby cover provides some protection should a sparrowhawk appear. Younger birds, kept from favoured feeders by the presence of the more dominant adult great tits, are forced to use those feeders that are located in more exposed locations, where the risk of predation is higher. This implies that the great tits are able to assess the risks and to preferentially feed closer to cover where they can. For the younger birds, the choice may be to feed somewhere that carries a higher risk of predation or to not feed at all: the balance of risks between predation and starvation.

While the sparrowhawk is an active predator, visiting the garden to feed on small birds, another bird of prey has been increasing its use of gardens in a very different way. The red kite is a large bird of prey, similar in size to a buzzard, and one that – like the buzzard – has seen a dramatic change of fortunes and a recovery in its breeding population. The red kite was lost from England and Scotland sometime prior to the 1880s because of persecution, leaving a small remnant population in Wales breeding in less than ideal habitats. Efforts to re-establish the red kite across the UK began in 1989, using wild caught young from donor populations elsewhere, since when the population has grown and spread to re-occupy many of its former haunts. This is one of the best examples of a well-researched and well-delivered reintroduction programme, a model for others to follow.

The kite is a scavenger, its diet dominated by scraps and carcasses gleaned from roadsides, refuse tips and other sites. Red kites have done well in cities like Reading, which lies close to one of the original reintroduction sites. In addition to the scavenged food, red kites are also taking food that has been deliberately provided for them at garden feeding stations. Many garden birdwatchers are, it seems, keen to see such a large and majestic bird swoop down into their garden to take meat and kitchen scraps put out for them. Some concerns have been raised about the provision of this supplementary food, in part because it is often of low nutritional value – it lacks bone and skin (which can lead to calcium deficiency) – and because most processed meats carry additives, and we simply do not know what impacts these may have on the birds.

One other aspect of red kite behaviour is worth a brief mention here and that is the preference that these birds show for adorning their nests with human materials, such as sheets, plastic bags and, at least on one occasion, cuddly toys. It is thought that the birds incorporate this material into the nest as a signal to other individuals. Work on the closely related black kite – a species in which roughly three-quarters of nests contain such material – found that the material is used to signal the dominance of the pair whose nest it is, presumably deterring other individuals from interfering with the nesting attempt. The behaviour is not new; it was referenced by Shakespeare in his play *The Winter's Tale* through the line '*My traffic is sheets; when the kite builds, look to lesser linen.*' The increase in red kite populations, both here in East Anglia and more widely across the UK, will no doubt see the bird return to popular literature, a lost cultural touchstone now reinstated.

* * *

Mid-month and there is a run of days that feels more like early autumn than early winter, the temperatures elevated by a succession of fronts pushing in from the west, each pulling in warmer air from the south. The flowering ivy remains surprisingly busy with late-season insects, the majority of which are wasps, and I wonder if the frost, when it comes, will kill these off. The warmer temperatures, pushing 17°C on some days, are accompanied by bands of rain and high winds. The combination of these conditions delivers some stunning skies, especially during late afternoon as the sun dips towards the horizon. On one afternoon the light from the sun illuminates some nearby poplars, bronzing their now bare branches against the backdrop of threatening clouds full of rain. The image reminds me of a scene in a book from my childhood, possibly *What to Look for in Winter* in the Ladybird series. A tinge of nostalgia touches me and I wonder how today's landscape might stand up to that portrayed in the paintings running through the book, and others in the series. I can immediately think of some of the most pronounced changes: the loss of our elms, for example, and the end of winter stubbles. Our bird populations have fallen by something like nineteen million individuals since the 1960s, and the structure of our bird communities has changed. One of the most notable changes is the increase in woodpigeon populations, one species that has clearly benefited from changing agricultural practices. Since the 1960s, its population has increased by 162 per cent; little wonder then that the species is now such a common sight at garden feeding stations, including my own.

One morning, with the temperatures dipping a little overnight, the dawn reveals thick fog and somewhere in the distance the calls of a flock of pink-footed geese. Newly arrived from Greenland, these birds should be feeding on arable fields on the coast, just a dozen miles north of here. Perhaps leaving their saltmarsh roost they have become disoriented by the fog, straying farther inland than intended. In a small way it hints at the challenges these birds face on migration, and their reliance on favourable conditions to ensure safe navigation towards their intended destination. These geese face other challenges. During the winter months they gather in vast flocks to feed on sugar beet tops in fields along the north Norfolk coast but, increasingly, the tops are now quickly ploughed back into the soil ahead of the planned next crop. In addition, the birds can damage winter wheat and so landowners use gas guns to scare the geese from their fields. The changing economics of farmland could spell disaster for these wintering birds, which may find themselves with limited feeding opportunities.

The fog has cleared somewhat by the time I venture out for a run. Passing the rookery, I am greeted by a chorus of cawing birds, perched high in the trees above me. Many of the birds are settled at or close to their nests, a marker of ownership ahead of the start of their breeding season, which is now only a couple of months away. Further on into my run and I pick up a starling in full song, perched on a chimney pot alongside a pair of jackdaws. It seems that the spell of mild weather has tempted him into song, alone on this street, the only of his kind greeting the day with his richly diverse notes.

The mild conditions make a noticeable impact on the bird feeders, whose contents sit largely unused as our garden birds continue to take advantage of the natural food supplies available through garden shrubs and trees, and others in scrub and hedgerows across the wider landscape.

It won't take much to see the birds return and for feeder use to jump again; one harsh frost could be enough to see things change. A few nights later and that frost comes as the temperatures drop below freezing; the first proper frost of the year. Come morning, the garden and its vegetation are crisp and white; the grass crunches under foot and holds my footsteps long after I have passed. Sure enough the birds return, as if a switch has been flipped, and I am conscious of the large numbers of blackbirds feeding on the windfall apples, and the tits and finches now busy at the feeders. The blackbirds are likely to be immigrants, individuals that have arrived over recent weeks and are only now moving out of the countryside as feeding opportunities there are diminished.

* * *

It is in November that the first of our wintering blackcaps puts in an appearance at the garden feeders. It is a male, recognisable by the black cap after which the species is named. Adult females and young birds of both sexes sport a brown cap. The bird continues to visit for several weeks, sometimes dominating the suet block and reacting aggressively towards the other small birds. The presence of the blackcap is a surprising one, but only because the species is traditionally regarded as a summer visitor to Britain, arriving here in May and departing in September. It is certainly true that the majority of our breeding blackcaps leave Britain in the autumn to winter in southern Europe and North Africa, at the western end of the Mediterranean. Historically, published county natural histories and the correspondence of country parsons make occasional reference to wintering blackcaps, but their presence here in winter is a more recent phenomenon. The numbers wintering here have increased substantially in recent decades, but it is only very recently that we have discovered why.

As work investigating the migration behaviour of European blackcap populations has developed, so we have come to understand more about the different strategies used. Blackcaps breeding in central Europe tend to migrate either southeast or southwest for the winter, the two strategies separated by a migratory divide running through eastern Germany, western Poland, the Czech Republic and Austria. A very narrow zone exists between these two populations within which birds show an intermediate strategy and migrate south. That this zone is so narrow suggests there is a strong selection pressure on these blackcap populations to avoid this route and instead migrate either southeast or southwest. Since the 1960s, a new migration strategy has evolved, with birds from continental Europe migrating north to Scandinavia, Britain and Ireland during the autumn.

A study published in 2020 revealed even more detail about the origins of those blackcaps wintering here in Britain and Ireland. By using tiny tracking devices, known as geolocators, the researchers behind the study were able to establish where the birds went after being tagged. The results revealed that our wintering blackcaps were drawn from breeding populations that extend across a 2,000-kilometre range, stretching from southern Poland to northern Spain. This wide geographic origin contrasts with the much tighter ranges of those blackcap populations following the traditional strategies of migrating either southeast or southwest on either side of that migratory divide.

Because of a changing climate and the increased provision of suitable foodstuffs at garden feeding stations here in Britain and Ireland, the blackcaps arriving here overwinter successfully and return to their breeding grounds before those individuals that have migrated far to the south. Because they return first, they pair with other individuals that have done the same thing. This favours the growth of this new migration route to Britain and Ireland, and the result is that we now see a growing wintering population. Examination of the citizen science data collected by BTO Garden BirdWatchers has revealed the importance of local conditions during the winter months. Wintering blackcaps are more likely to use gardens in cold winters, and to favour gardens located in the warmer parts of the UK, such as southwest England. In addition, they are also more likely to use those gardens where the provision of food is regular, underlining the importance of a dependable source of food.

The reliability of a food resource is important for other small birds too. While the blackcap visits to feed, some species seek to increase the food resources available to them by adopting other behaviours. These can include food defence, something that some of the blackcaps certainly indulge in, and food caching or hoarding. This latter behaviour is behind the frequent visits we have seen on recent mornings by one or more coal tits. These come to the feeder, avoiding the larger and more dominant species, take a sunflower seed and then disappear. While some of the seeds may be eaten elsewhere, others are cached, pushed into a crevice or other site to be eaten at a later date. That some of the seeds are cached becomes clear the following year when sunflower seedlings emerge from the moss on the roof.

Another species renowned for its caching behaviour is the jay, a bird that may collect and cache as many as 5,000 acorns during the window of their availability. The scale of the acorn crop can also drive the movements of jays, with birds moving over significant distances if they are unable to find acorns locally. Such movements can see an autumn influx of jays into the UK from the Continent. Interestingly, jays may alter their acorn caching behaviour if there are other jays around. Fearing that their caches may be found by other individuals, they increase the distance travelled between where an acorn is collected and where it is hidden. Quite how the jays and coal tits remember where their acorns and seeds have been cached is unclear, and many nuts and seeds do go uncollected. Some are retrieved within a few hours, providing a temporary buffer should feeding conditions change during the course of a day, while others are left for many days or even weeks. Laboratory studies have revealed that an individual coal tit's ability to remember where it has cached seed lasts for less than six weeks. Because coal tits tend to cache seed in the types of site where they would normally search for food, this suggests that the choice of site for caching provides a mechanism by which a coal tit can increase its chances of finding the stored seed at a future date. Of course, food caches are going to be more important for a coal tit living in a forest or woodland, where changing weather conditions can alter the availability or chances of finding food, than it is for one living in a sprawling suburb. However, the behaviour remains, even though the hanging feeders and bird table ensure a regular supply of suitable seed.

While the coal tit's small size may place it low down the pecking order at garden feeders, it does have one advantage when it comes to finding food. Its small body size makes the coal tit an agile feeder, able to forage at the ends of conifer branches where a larger bird would struggle. Although hard to observe from the ground, coal tits spend a lot of time feeding in the upper reaches of large conifers, hopping and fluttering between the branch ends and cones in search of insects and spiders. Prey is found by active searching, the tit minutely examining surfaces, probing between pine needles or dislodging lichen from larger branches as it moves about the canopy. On occasion, a coal tit will hover to extract a seed from a part-open cone or to pluck a spider from the tip of a branch that is too small and delicate even for a bird of this size. Only the tiny firecrest and goldcrest can gain similar purchase here.

The shortening daylight hours of late November see increased activity at the feeding station. The birds have less time in which to feed and their numbers have been increased through arrivals from farther afield. A growing flock of chaffinches can be seen feeding on the ground below the bird table most days now. Perhaps some of these birds will have been feeding on beechmast over recent weeks, only moving into the garden now that the beech crop has been diminished. The flock contains a mix of male and female birds, some of which have legs encrusted with powdery-white lesions. In some cases these are small, affecting only a small part of the leg or foot, but there are at least a couple of individuals showing a significant growth over much of the leg. These lesions are the external signs of disease, associated with either a virus, a mite or both.

The one that many birdwatchers will have heard of is chaffinch papillomavirus, sometimes called 'tassel foot', which appears to be endemic in the chaffinch population; the other is caused by tiny mites, which burrow under the leg scales. The skin abnormalities, in the form of warty lesions and growths, appear to develop slowly over a period of weeks or even months. In some cases, the impacts are restricted to a small and discrete lesion, somewhere on the leg or foot. In other cases, a much larger growth can form, perhaps even resulting in lameness or the loss of one or more toes. In the worst cases, a finch may lose its entire foot. Diagnosis of the two conditions requires laboratory tests, and research work here in the UK indicates that some birds may be suffering from both at the same time. Studies from across Europe seem to indicate that roughly one in five individuals in a flock may be suffering from one or other of the two diseases, but we have yet to see an epidemic affecting a much greater proportion of the population.

The sight of the diseased chaffinches is a further reminder, if one is needed, of the disease risks faced by wild birds and the responsibility placed on us to ensure that we keep our feeders and bird tables clean and disease free. The risk of disease transmission is naturally greater where large numbers of individuals gather together, something of which we have become very much aware since the COVID-19 pandemic. The same is true for wild birds, with the risk of transmission greatest where birds either roost or feed communally. A garden feeding station, contaminated by droppings or by spilt food, provides an opportunity for disease transmission and for this reason it should be cleaned regularly with a suitable detergent. These finches are feeding on the ground, so I make a mental note to move the feeders and bird table to a new patch of ground.

* * *

Away from the bird table and hanging feeders the birds are also feeding on the berries that remain in the hedge and on the shrubs. It has not been a particularly good year for berries, the weather conditions conspiring to prevent the development of a large crop. With the exception of the ivy, which is only now coming into crop, there is little left for the visiting birds. The holly still holds a few red berries, which for a day or two are defended by a visiting mistle thrush. Food defence of this kind is really only worthwhile if the food is scarce elsewhere and, where found, can be easily defended. In the past, I have seen both fieldfare and mistle thrush attempting to defend a berry-producing shrub from a horde of other thrushes. Valliant though the defence may be, the defender is invariably overwhelmed by the number of birds keen to feed on a diminishing resource. I ponder on putting out some of the windfall apples that we have stored, but I do not have too many of these this year, so I decide to keep them back for later in the winter. If we get snow, then they are likely to be of more value then.

I wonder what the berry crop has been like farther south this year. It is known that berries can be an important food for migrating thrushes and warblers, their nutritional composition ideal for fuelling migration and enabling birds to refuel quickly at stopover sites on the journey south. Such stopover sites enable migrating birds to break their journey into a series of shorter flights, the birds able to rest and refuel between periods of sustained flight. Not every species breaks their migration into several discrete sections; it is dependent on how far they have to fly in order to reach their wintering grounds, what sort of terrain they have to cross, and how quickly they need to move. Of course, some small birds are able to feed on the wing, so may not need to break the journey to feed, though they may break the journey to rest or to roost if they only travel at certain times of the day.

I also wonder about those birds that were here just a few weeks ago: the late chiffchaff that passed through in September, the pair of blackcaps that came to the pond in summer, and the swifts that were overhead during June and July. Where are these individuals now? Did they complete their migration successfully? There's a chance that both the blackcaps and the chiffchaff are wintering somewhere in southern Europe, close to the Mediterranean. Perhaps they have even crossed into North Africa to winter on the same continent as the swifts, who will by now be somewhere in central or southern Africa. I wonder who else has seen these birds, either on their journey south or now that they have reached their wintering areas, and it reminds me that our migrant birds do not truly belong to us; they belong to all of the communities whose lives they touch. While this is particularly true of species like swift and cuckoo, who spend such a tiny fraction of their year with us, it is also true of all of the birds who pass through here and visit elsewhere.

One UK bird that is very far from home, or at least its original home, is the ring-necked parakeet. I have seen a single bird here in Norfolk, but I see them fairly often when I visit London for meetings. Preferring to walk from King's Cross, rather than get a tube, I often see and hear them as I walk through the green squares or urban parks for which London is rightly known. London and the Home Counties now support a sizeable population of this introduced parrot, whose origins lie on the Indian subcontinent but whose addition to our avifauna has its origins in the pet trade and aviary escapes. These small green parrots have been breeding in the wild in the UK since at least the late 1960s, and while we have seen their numbers increase markedly since this time, we have yet to see a substantial colonisation of a wider area. Ring-necked parakeets form large communal roosts at favoured sites, numbering many hundreds or even thousands of birds, and it may be this behaviour that has, so far, prevented a significant expansion in the bird's distribution across the southeast of England.

One of the problems with the parakeet as a garden bird is that flocks can dominate garden feeding stations, damaging bird feeders and reducing feeding opportunities for other species. While the parakeets compete for food they can also intimidate other garden birds, such that the mere presence of a flock in a garden can deter other birds from coming in to feed. Concerns have also been expressed regarding the parakeet's potential impact on breeding species, like starling and nuthatch. Both are cavity nesters and so compete with the parakeets for suitable sites. The parakeets can be aggressive and seem well able to turf other species out of chosen nest cavities. Although there is no evidence yet for any significant impact here in the UK, work elsewhere has found that the presence of breeding parakeets reduces the local abundance of breeding nuthatches. This is certainly one to watch, and a species that I am not keen to see arrive at my garden feeding station, despite the splash of colour that it would bring to dull winter days.

Ring-necked parakeets are not the only exotic birds to turn up at garden feeding stations. From time to time over the years I have been sent photographs of unusual birds, photographed by garden birdwatchers unable to put a name to their visitor. Most of these turn out to be aviary escapes, birds kept in captivity that could not possibly have made their way to the UK from breeding ranges located in southeast Asia, Africa or South America. As you might expect, many of these aviary escapes have colourful plumage and most are seed-eaters, hence arriving at garden feeding stations when suddenly finding themselves out of an aviary and faced with the challenges of finding food in the wild. Larger, more unusual, birds turn up from time to time, including escaped eagle owls, vultures, hawks and members of the pheasant family. One of the strangest I encountered through a photograph was an intermediate egret that walked into a conservatory and promptly went to sleep. I have certainly never had anything more exotic visit my garden than an escaped budgerigar.

* * *

The second half of the month sees a continued upturn in the numbers of birds visiting the feeders, with a growing flock of ground-feeding finches and blackbirds, the former to a scatter of seed and the latter to a few windfall apples that are delivered courtesy of a friend. While the garden feeding station is undeniably busy, there still seem to be plenty of birds active in the surrounding countryside. Morning walks out along the old railway line see us pass large flocks of fieldfare working the fields. On one morning, the fieldfares are scattered through a flock of larger birds, a mix of lapwings and golden plovers. The following day, driving along a country lane that borders one of the large country estates, a mixed flock of chaffinches and bramblings lifts from the road ahead of me. The birds have been feeding on beechmast, presumably broken open by passing cars, and when I return an hour or so later I slow to a halt well ahead of the roadside beech trees under which the birds have been feeding. Pulling off the road, I can stop and watch the flock through the binoculars that I keep in the car for just such an occasion. It is wonderful to see the bramblings, the males in particular more strongly marked and richly coloured than the chaffinches alongside which they are feeding. I wonder how long the beechmast will last, and where these birds will go next. I hope that some, at least, might visit my garden.

It is this balance between the food available within the wider countryside, the prevailing weather conditions, and the offerings provided through my garden feeding station, that is central to how busy my feeders and bird table are through the winter months. Harsh weather on the Continent can force birds to push farther west, perhaps delivering a pulse of new arrivals overnight. Add to this the variability in size of natural seed and berry crops and you get a sense of how birds may be forced to respond to conditions over large areas. A combination of poor weather and poor crops can lead to major movements, bringing large numbers of finches and thrushes into the UK and its gardens. In contrast, a mild winter and an abundant crop of berries and seeds close to the breeding grounds will see these same birds remain where they are, leaving UK gardens relatively bird free.

Another factor to add into this mix is changes in the management of natural and semi-natural habitats. The creation of new forests, perhaps established on a commercial basis for the production of softwood timber and paper, might increase the availability of conifer seeds as the trees mature, enabling an expansion in siskin populations or a change in their seasonal movements. This might reduce the numbers visiting garden feeding stations, or perhaps even have the opposite effect as the siskin population increases because of the new habitat being created. The same pattern may result from changes in agricultural practices, such as the shift from spring to winter cereals that reduces the availability of overwinter stubbles. In this case farmland buntings and finches, which would have fed on the stubbles, increasingly turn to garden feeding stations during the late winter months because there is little seed left in the wider countryside. Changing management practices may be a consequence of changing economic practices or policy decisions; they may also be a consequence of a changing climate, an area perhaps no longer able to sustain a particular management approach under the new environmental conditions.

Whatever the cause, it is clear that bird populations exist in a state of flux and are often able to move large distances to locate better feeding opportunities elsewhere. While this might suggest a certain resilience in their populations, it is clear from the massive long-term declines seen in many of our breeding populations that all is not well; that the changes we are seeing now in our countryside are having a tremendous impact on populations of wild birds. While we might be delighted to see a new bird in the garden, it is worth pondering on the reasons for its arrival, especially when viewed in the context of what is happening elsewhere. It is perhaps for this reason that I tend to take more delight in the common species, the residents that are here throughout the year. These are the individuals with which I share the garden and its resources. It is as much their garden as it is mine, and it is my duty to ensure that I manage it in a way that complements their lives. From the timing of when I trim the hedge or cut the lawn, through to the branches that I place in the pond or the food that I provide on the bird table, I want my actions to be a positive contribution to those of my avian neighbours.

December

Late one morning, the first weekend of the month, I spot three small thrushes perched in the rowan that I planted a few years ago; I remember how it only just fitted in the car, so keen was I to have this valuable berry-producing shrub in the garden. The thrushes are redwings, my first in the garden this winter but hopefully not the last. Rowan carries clusters of small white flowers from late spring, which are valuable for visiting insects, but it is the orange-red berries that are important for birds and the reason why I wished to add this shrub to the garden. I'd wanted the native form of the shrub rather than one of the ornamental varieties like 'Sheerwater Seedling' or 'Xanthocarpa', the latter with its pale orange/yellow fruits. Unfortunately for these redwings they are too late, the crop of berries having been finished by visiting blackbirds several weeks earlier.

Many plants rely on birds to help disperse their seeds. The nutritious fleshy fruits and berries that contain the seeds are offered by the plant as incentive, luring a bird to take the fruit, and the seeds hidden inside, and to carry them away elsewhere. Each seed needs to have a tough external coat if it is to survive the bird's digestive system and pass through unharmed. This apparently mutualistic arrangement (the plant offering some form of reward, the bird acting as a dispersal agent) works well, though there are cheats in the form of birds that either eat the pulp but discard the seed below the parent plant, or eat and digest the seed itself. Different fruits and berries may become available at different times of the year, and some will remain on the plant for longer than others. Holly is a good example of a shrub where the berries can remain on the tree for a substantial period. The timing of fruit production can even vary across closely related species. Rowan (*Sorbus aucuparia*) ripens from as early as late July, while whitebeam (*Sorbus aria*) ripens in September, and wild service tree (*Sorbus torminalis*) from late October. In this particular case, the differences may be linked to geographic range: rowan is the most northerly distributed of the three shrubs, and those shrubs with more northerly distributions tend to fruit earlier than those from farther south. This may act as another incentive for berry-eating birds to move south with the winter; if they stay in the north, not only will they have worsening weather conditions to contend with, but also a declining supply of their favoured food.

In addition to differences in the timing of their availability, berries may also differ in the nutritional rewards that they offer. As a berry ripens and matures, its chemical composition changes, typically with the water content of the berry's pulp declining and its lipid content increasing. Lipid content is important for berry-eating birds, not least because, once metabolised, lipids produce more energy (per gram of dry weight) than either carbohydrates or proteins (the other main components of berry pulp). It makes sense for birds to select those berries that offer the greatest nutritional returns, something that may be influenced not just by the composition of the berries but also their abundance and ease of access. A rowan with an abundance of ripe berries might be the best option available locally in terms of nutritional rewards, but if it is being actively defended by a fieldfare or mistle thrush it suddenly becomes a less attractive option for a smaller thrush, like a redwing or blackbird.

Evidence from scientific studies highlights how birds use the colour of a berry to assess its nutritional value. Many berries are black or red, some are orange or yellow, but (at least in the UK) very few are white. The colour of these fruits is determined by pigments, including the anthocyanins, which are black- or ultraviolet-reflecting. Anthocyanins are known antioxidants, so there may be additional nutritional rewards for those birds actively selecting fruits that have a high anthocyanin content; in which case the dark colour of berries rich in anthocyanins may be an honest signal of the rewards on offer. Watch the berry preferences of visiting birds and you might see for yourself how black and red berries are favoured over those that are orange, yellow or cream. This may be particularly obvious if you just look across a small group of related shrubs, such as the *Sorbus*, for which there are many different ornamental cultivars producing berries that range from creamy white and yellow through to the deepest red. It makes sense that ornamental fruits, whose colours are not widely replicated within native species, are likely to prove less attractive to visiting birds – another reason to purchase and plant native shrubs in your garden.

Other factors can be of importance when it comes to fruit selection, a key one being what a particular species is looking for. Greenfinch is a seed predator and takes a broader range of seeds than probably any other finch in the region, so the nature of the pulp is likely to be less important in shaping its decision about which fruits to target. This familiar finch readily takes to ornamental shrubs introduced into gardens, especially the hips of roses, and the berries of *Sorbus* cultivars and cotoneaster. Despite their toxicity to mammals, the arils of yew and their seeds are also targeted by greenfinches. Other species of bird are after the pulp; some, such as blackbird, are fairly catholic in their tastes and take a range of fruits and haws, while others show some clear preferences. Song thrush favours elder, guelder-rose, the sloes of blackthorn and the arils of yew. The larger and tougher fruits can be a challenge for smaller thrushes, so only fieldfare, mistle thrush and to a certain extent blackbird can cope with rosehips, largely leaving song thrush and redwing to the other smaller fruits.

What all this means for the garden birdwatcher, keen to encourage a wide range of birds into their garden, is that the types of berry-producing shrubs chosen for a wildlife garden should be selected with care. It is all too easy to be tempted by an unusual colour form of a familiar species, but will it deliver for wildlife? And if you want to attract redwings, fieldfares or waxwings you need to understand their feeding preferences so you can plant the types of berry-producing shrubs and trees that they favour. Waxwing is an occasional garden visitor, whose numbers can vary substantially from one year to the next. The reason for this lies in the waxwing's winter diet, which is dominated by berries; it has been estimated that a waxwing may consume between two and three times its own bodyweight in berries each day. The waxwing's diet may also be behind the large quantities of water these birds imbibe. Because a diet of berries is particularly carbohydrate-rich, this needs to be balanced by large quantities of fresh water. A study of the closely related cedar waxwing – found in North

America – revealed that each bird excreted at least double the amount of water available in its body each day, underlining the volume needed to replace that being lost, some 42 millilitres. The water demands of waxwings can sometimes get them into difficulties, however. Where the birds are forced to drink from puddles formed on roads that have been gritted with salt, they can become intoxicated and die, something witnessed in Dresden city centre in 2005, where five per cent of a 600-strong flock was found dead or dying because of salt poisoning.

Although the species regularly overwinters within its northern European breeding range, it is prone to make large-scale eruptive movements when berry crops have been poor. These movements can bring large numbers to the UK, where the birds can often be found in urban areas, favouring the berry-producing shrubs characteristic of amenity planting on industrial estates, supermarket car parks and new housing estates. Small groups of waxwings, sitting in the tops of small trees, are reminiscent of starlings in terms of body size and outline, so any such group is worthy of a second glance during one of those winters when large numbers are known to be in the country. These are surprisingly confiding and approachable birds, perhaps reflecting the remoteness of their sub-Arctic and Boreal zone breeding grounds. Feeding groups can provide excellent opportunities for birdwatching or photography, interspersing these bouts with periods during which the birds loaf around while digesting their meal. If you don't have suitable berry-producing shrubs in your garden, it is worth knowing that waxwings can be tempted to feed on apples, so long as they are still attached to the tree. Since you are unlikely to have any apples left on a tree by this time of the year, the solution is to cut some in half and then tie them to the branches of any suitable tree, or hang them like Christmas decorations. Judging from photographs posted online, this trick has worked particularly well for the residents of Shetland and Orkney – two of the first places to receive visiting waxwings in those years when they arrive. Recently, there was even a photograph of two young men, quietly sitting with backs to a fence while caressing apples in their hands. Each had two or three waxwings perched upon their hands, legs or wellington boots, the birds taking turns to feed from the apples. Human apple trees – now there is an interesting approach to feeding garden birds!

* * *

While most of the birds visiting the garden in December are likely to be foraging over a wider area, one or two species will be settled on a winter territory and less mobile in their habits. Perhaps the most obvious of these is the robin, whose winter song is one of the few to be heard at this time of the year. To my ear, and indeed to those of many others, the robin's winter notes are more melancholy in tone than the phrases that make up the breeding season song. Although different in character they serve a similar purpose, advertising ownership of a territory that will be defended against intruders. Establishing a winter territory has several advantages for a small bird, the most important of which is the defence of a suitable food resource. Studies of wintering robins have revealed that competition for winter territories can be fierce, and that territory-holding individuals are more likely to survive the winter than those who fail to secure one.

In some species, such as wren and chiffchaff, these winter territories are often located in sites where conditions tend to be warmer and damper than elsewhere, underlining that these are exactly the sorts of places where insects and other invertebrates may continue to remain active in all but the coldest weather. Reedbeds and areas of wet woodland appear to be a particular favourite with wrens, and indeed with the small numbers of chiffchaffs that now overwinter in southern England. It is not just the access to food that a territory provides that is important; ownership of a territory and its use over many weeks brings with it knowledge of where that food is most easily found. Such knowledge can be particularly valuable when conditions are difficult. This is something that has been well-studied in the tawny owl, the ultimate in sedentary, year-round territory-holding birds. The 'local knowledge' that comes from long-term ownership of a territory has been found to be particularly important to breeding tawny owls, with established pairs able to breed successfully in poor years because of this, while inexperienced birds – recently established – struggle to find food and are more likely to fail in their breeding attempt.

Given that such winter territories are structured by the availability of food, it is unsurprising that it is not just the males who establish them. Many female robins will establish a small winter territory close to where they will breed the following year, but a few are forced to move farther afield. Even where they winter some distance from their breeding territory, most females will return to their original territory and partner once winter comes to an end. Most of the robins hatched in the UK will spend the winter within a short distance of where they were raised, but a few may winter as far south as Spain.

However, it is the local movements that are of greatest interest to the garden birdwatcher, since it is these that may bring robins into gardens from other habitats. You only have to look at the BTO Garden BirdWatch weekly reporting rate graphs to see this, with the use of gardens far greater in the winter months than during the breeding season. This reminds us that while many gardens may be unsuitable for a breeding pair, they are able to support individual birds during the winter months.

That a garden can support a robin during the winter months but not during the breeding season demonstrates that birds may require alternative resources and have different needs at different times of the year. For most garden birds, the breeding season diet is dominated by insects and other invertebrates, particularly so for chicks in the nest, but outside the breeding season the diet is typically more varied. In great tits, for example, the autumn and winter diet is dominated by seeds, and subtle seasonal changes in the shape of the bird's bill reinforce this switch away from soft-bodied invertebrates to harder seeds. The pattern is repeated in thrushes and finches, the birds responding to the changing abundance of particular food types. There are other changes too; outside of the breeding season, birds are no longer tied to a nest site and a breeding territory but can range more widely. As we have just noted, some species may switch a breeding territory for a winter territory; other species do not settle in one place but instead range widely across the countryside, perhaps even crossing borders in the process.

A winter garden offers food – the main reason many of the birds have for visiting – and roosting opportunities. We know that food is a particularly important driver of garden use, because we have been able to examine the data collected by BTO Garden BirdWatch participants and determine which of the factors offered by a garden (such as the number of tall trees present, the availability of berry-producing shrubs, and the presence of food, etc.) is the most important. For all of the garden bird species studied, food was a significant factor in determining whether or not a garden was used. Given the scale of food provision that we noted back at the start of the book – the £200 million spent in the UK annually on bird food and bird care products – and the reliability of that provision, it is little wonder that our wintering bird populations make such good use of it.

* * *

The brief sighting of a fox in a nearby garden, reported by one of our neighbours, comes as a surprise. We were not aware of their presence locally, this far into the town, and the species is persecuted across much of the nearby countryside because of game-rearing interests. Even with the persecution, the foxes still manage to maintain a population here, living under the radar, their presence only revealed by the strong scent that is sometimes caught on the wind during a walk along the old railway line out of town. One morning, back in the summer, we watched a dog fox cross a field of sugar beet and pad its way down the slope and into the thick cover of the wooded strip that masks the permissive path back to town. It was the first that we had seen here. Urban fox populations appear to have increased, supported by the quantities of waste food that can be found in most towns and cities, and I wonder if we might see more individuals in the future.

To hear of one sighted close to the garden delivers a sense of connection to the wider countryside and takes me back to my childhood garden on the Surrey–Hampshire border, where a fox was a regular visitor and one that was encouraged by scraps from our dinner table. The animal kept fairly regular hours and would visit our feeding station during the early evening, its appearance revealed by the security light triggering. The fox would investigate the scraps offered in an old pie dish, select what it wanted and trot away with a mouth full of food. Other visits were almost certainly made later into the night, the pie dish usually discovered the next morning some distance away from where it had been placed the evening before. On one occasion we found an old shoe left in its place – foxes have a habit of scavenging items that they deposit elsewhere – which left us wondering what its original owner might have made of its disappearance.

Many garden birdwatchers extend their food provision beyond birds to provide for foxes, while others feed badgers, and a very fortunate few receive visits from one of our rarest mammals, the pine marten. Some, including our nearest neighbours, set up a hedgehog feeding station – a dish of meat-based cat biscuits placed under a large plastic box into which a hedgehog-sized hole has been cut. This is well used and likely to be an important source of food during the dry summer months when favoured invertebrates are difficult to find. Feeding mammals is no different from feeding birds. However, the nocturnal habits of many mammalian visitors means there is often less opportunity to watch them visit and feed, which is why camera traps have become so popular.

* * *

Just before Christmas, I notice that one of the dunnocks visiting the bird table has a small warty growth beneath its eye. At first glance it looks like a tick, the rusty colour of the wart similar to that of a newly attached tick, but once seen through the binoculars it is clear that this is not the case. These sorts of small wart are typically the result of an avian pox virus, something that appears to be endemic in the UK and not uncommonly seen in dunnock, house sparrow, blackbird, starling and a handful of other species. The virus causes these discrete warty lesions on the featherless parts of the bird, typically around the eyes, on the legs or feet. Most are small in size and do not appear to hamper the individual affected. Evidence suggests that most such infections last for less than a few months. However, more significant lesions are sometimes seen and this appears to be typical for the pox lesions noted over recent years in UK populations of blue tit and great tit.

The emergence of these more significant lesions in tit populations seems to have its origins in 2006, when cases were first reported from the south of England. The pattern of cases reported through Garden Wildlife Health, a citizen science scheme monitoring disease in garden wildlife, indicated that the virus may have arrived on England's southern shore from overseas, most likely via an infected biting insect blown across on a warm plume of air. As well as via biting insects, the virus can be transmitted through contact between individuals and via contaminated surfaces. Genetic examination of the virus recovered from some of the infected individuals showed it to be almost identical to that documented elsewhere in Europe, supporting the hypothesis that it had arrived via the south coast through a biting insect – there is no interchange in great tit and blue tit populations between the UK and the Continent, the birds being resident in habits. The virus was then seen to spread northwards across England, reaching both Wales and Scotland.

While reports of avian pox in other species, such as my visiting dunnock, tend to refer to just a single individual with symptoms, a characteristic of the infection in great tit and blue tit populations was the large number of individuals being seen together by garden birdwatchers. These garden birdwatchers were not just concerned by the numbers of individuals affected but also by the severity of some of the infections. People were seeing great tits with lesions the size of a marble on their head, in some cases seemingly preventing the bird from seeing out of one eye. In the worst cases, it appeared that individuals had died because their vision was so hindered that they had flown into windows or other objects. Work on a study population of great tits breeding in Wytham Wood just outside Oxford, which has been going on for many decades, charted the arrival of the disease and its impact on the tits breeding within the study population. Infected individuals showed lower levels of breeding success, linked to a reduction in their ability to find food and to feed their chicks. There was evidence of transmission from infected adults to their chicks, a reduction in survival rates and a growth in the proportion of the population showing signs of disease. Within two years, some 10 per cent of the 8,000-strong population had been infected. The longer-term impacts of this form of avian pox virus are unclear but it is worth noting that, following a long period of population growth, the UK great tit population has now fallen by 12 per cent over the last ten years. This is something that will require further study.

I have yet to see Paridae pox – as the disease in great tits and blue tits is known – here in the birds visiting the garden, so we have been fortunate. Again, it is a reminder of the need to maintain good hygiene at the feeding station, and to be alert to any potential problems that might emerge. The dunnock continues to visit over the Christmas period and there is no sign of any change in the size of the lesion below its eye. It seems to be feeding normally and there is no indication that the pox lesion is having any impact on the bird's daily life.

* * *

The week between Christmas and New Year is a relaxing one, spent at home and either in the garden or out locally for a walk. Late one morning, just before a planned early lunch, I spot a marsh tit on the bird table – the first one to have visited this garden. The identification of this individual, about the size of a blue tit but with grey-brown upperparts, buff underparts and a glossy black cap, is not straightforward. In appearance, it is similar to the closely related willow tit, the two species differing only in a small number of subtle plumage features – and their song and calls. So similar are the two that the willow tit has the distinction of being the last regularly breeding British bird to be identified and named. Although the willow tit had been described as new to science in 1824, it wasn't until 1897 that its presence in the UK was revealed by two German ornithologists who had been examining a series of marsh tit skins in the British Museum's collection. The ornithologists spotted that some of the marsh tit skins were in fact those of willow tit. Three years later, the species was officially added to the British list.

Sadly, both species have undergone significant declines in the UK, with the marsh tit population falling by 80 per cent between 1967 and 2017, and the willow tit population by 93 per cent over the same period. We have now just about lost willow tits from Norfolk, but the marsh tits are still here – albeit in reduced numbers. There is evidence that changes in woodland habitats, particularly the loss of understorey vegetation and a drying out of favoured damp sites, may be responsible for the decline evident in marsh tit populations. The same factors may be behind willow tit decline, but the evidence is less clear, partly because of a lack of research and suitable data sets.

The marsh tit doesn't stay long and nor does it put in another appearance, despite my increased vigilance. Adult marsh tits, settled in established pairs, tend to remain in their territories throughout the winter months, meaning they are unlikely to move farther afield and visit garden feeding stations. Young birds may remain attached to broader social groups, but some, perhaps one in three, may forage alone or attach to a mixed-species flock. I suspect that this was such an individual, making a brief foray away from the wet woodland on the edge of town with its small breeding marsh tit population. It has been estimated from studies of English marsh tit populations that an individual must spend 80 per cent of the daylight hours actively foraging for food during the mid-winter months, visiting 1,100 trees and needing to find one average-sized insect every minute in order to maintain itself. Such figures, although difficult to determine accurately, still provide an indication of the pressure that small birds are under if they are to survive the winter. When viewed against this, it is easy to see why the reliable supply of oil-rich seeds provided at a garden feeding station can make such a difference to overwinter survival rates.

* * *

As December edges towards its end, I begin to think of the year ahead: of the first buds, the first singing blackbird and, months later, the return of those summer migrants that, even now, are already beginning their journeys north from wintering grounds in the southern half of Africa. The swifts will be returning to Central Africa, somewhere near the Congo Basin, and come spring will then push west to Liberia, which appears to be an important stopover site for UK birds before they make a crossing of the Sahara. West Africa is an attractive stopover site for many of our spring migrants, something almost certainly linked to the flush of new growth, and associated boom in invertebrate life, that comes with the drought-busting rains that push north and west across the African continent at this time of the year. We see a similar pattern in our cuckoos, tracked from wintering grounds in Central Africa as they begin their return migration to our shores. These birds, the swifts and cuckoos, underline more than any other the degree of connection that exists between our gardens, the surrounding countryside and the distant wintering grounds. What happens to these birds on migration, perhaps at refuelling sites around the Mediterranean or in northern Italy, will determine if they make it to their wintering grounds, or indeed back to their breeding sites here in the UK. Migrant birds connect us to other communities, over vast distances, in a similar way that my winter-visiting great tits connect me to nearby woodland, and the overflying oystercatchers link me to wintering sites on the north Norfolk coast and breeding sites in the fields that surround the town where I live. It is quite something to bring each of these birds into focus and to ponder on where it is now and what conditions it is experiencing.

It is amazing to think of my garden bird community in these terms, a small part of a much wider web that reaches out to touch other landscapes, other avian communities and other people. The complexity of food webs is something that I learned about during my first term at university, but even those lectures failed to capture the interconnectedness that is present at a global scale. The sunflowers, whose seeds I pour into my bird feeders on a daily basis during these cold December days, would have been part of a food web where they were grown. Now their seeds are part of a food web here in my garden, and while they will not have fuelled the migratory journeys made by the blackcaps now wintering in North Africa, they may fuel the journeys that the chaffinches and bramblings using the feeders today will make in a few weeks as they return to breeding sites in Scandinavia. While the predatory sparrowhawk will undoubtedly take some of these wintering finches, it will be other predators that will attempt to capture them once they return north, where those that survive will seek out prey of their own, in the form of insects and spiders to feed to the nestlings they go on to raise.

It is these food webs that are central to functioning natural systems, systems that we know are changing because of the impacts of our activities on the landscape and climate. Changing land use alters feeding opportunities, increasing the availability of some resources but decreasing the availability of others. The food provided at my garden feeding station, and at thousands of others like it across the country, provides an additional resource for seed-eating birds; at the same time we have seen the loss of the woodlands, hedgerows, scrub and rough grasslands that would have provided more natural feeding opportunities for these birds. Does providing seed in hanging feeders and on bird tables replace such losses? To some extent it does, but does the scale of this supplementary resource make the loss of natural and semi-natural habitats acceptable? Almost certainly not.

We have seen the impacts that disease can have on bird populations and we know that the risk of disease transmission is increased where birds gather together in large numbers, such as at garden feeding stations. By removing feeding opportunities in the wider countryside we are perhaps crowding seed-eating bird populations into a smaller suite of feeding stations – albeit still many tens of thousands in number. Importantly, we are also seeing species feeding together in close proximity that almost certainly would not mix in this way under more natural conditions. But does this mean that we should stop feeding? No, it doesn't, but it does mean that we should think carefully about what and how we feed. In particular, it also means that we should pay more attention to hygiene and to how often we clean our feeders and bird tables. Our actions have consequences and we want them to be positive.

The food that we provide is an important resource, and one that does make a difference to those individual birds that choose to visit and feed. And this is one thing I have learned over the years, that each of the birds visiting is an individual. It might be an individual that makes just a single visit to one of my hanging feeders, never visiting the garden again, or it might be an individual hatched locally which will spend its entire life within a short distance of the garden. Whether it is a once-only visitor or a near neighbour, its fortunes are being shaped by my activities. The food, cover and nesting opportunities that I provide should be a positive influence; were I to use pesticides, replace the lawn with artificial turf or replace the hedge with a fence, then these actions would be likely to prove damaging. On the same basis, decisions about the food that I eat, how I travel and what I buy have consequences for other plants and animals, living in other habitats or in other countries. It is for this reason that I view my relationship with the wildlife in my garden in the way that I do. It is the model for how I view the wider natural world of which I, and every one of us, is a part. By recognising the birds in my garden as individuals, I am framing how I view the wider world. The respect that I hold for these individuals is matched by the respect I hold for creatures elsewhere, who I will never see or know. And that is why being able to watch birds in my garden provides not only a window onto their lives and behaviour, but also onto the natural world as a whole.

The garden bird year has its patterns, marking the change of the seasons and the succession of generations. From the challenging winter days, when birds crowd onto the bird table and feeders, to the slow days of late summer, when a new generation of young birds can be seen visiting, there is a sense of hope and renewal. Life goes on, and the pattern to the year's passing is reassuring. Things are working, they are not yet broken, and this comforts me. The English painter Stanley Spencer was once criticised for his parochialism and interest in the small things of everyday life. But it is through these things that you develop a more complete understanding of the wider domain, both human and natural. Attention to the detail of life within the garden provides a richness of understanding that both adds to our sense of well-being and enables us to better comprehend the pattern of life across a global stage. Understanding the harsh realities of predation, witnessed in graphic detail when a sparrowhawk makes a kill close to the bird table, frames my understanding of a pride of lions bringing down a wildebeest far better than any television documentary. The natural world is an amazing thing, full of complexity, and while my garden represents just the tiniest fraction of the wider whole, it reminds me of my place within the natural world because this is where I make my connections with individual birds and other creatures. It is the world in miniature, and with the passing of each year the importance of this grows inside of me.

Afterword

The 'garden' in this book is drawn from the three different gardens that I have owned, all located in Norfolk. The encounters with the birds and other wildlife documented in the twelve monthly chapters were captured in notebooks or held in the mind. Because each is focused on the encounter, I hope that there is little sense that these have come from different gardens. If you have seen through this small deception then I hope you will forgive me and that it has not lessened your enjoyment of this book. In my defence, virtually all of the encounters described could have happened in any of these three gardens. Over the years, I have learned that time spent watching brings its rewards, and the more that you watch the more you see. It is time that delivers that rare chance encounter, though helped by the feeding, roosting and nesting opportunities that you provide.

The first garden that I owned was the most urban, a small end-of-terrace plot less than 200 metres from the high street of a market town in southwest Norfolk. Despite its urban location it still attracted blackcaps each spring, redwing in the winter and the occasional lesser redpoll. Once it was visited by a muntjac, an animal that had somehow cleared a four-foot fence and which then had to be captured in a large dog crate so that it could be released from the garden. The second was rural, surrounded by farmland and rich in birdlife. Chiffchaffs bred just beyond the hedge, red kites and hobbies were regulars overhead in the summer, and chattering swallows sang from the wires coming into the house. My current house is urban, again located close to the centre of a market town and again part of an old terrace. It too has blackcaps visiting, swifts overhead in the summer, and a mix of finches and thrushes in winter. Each of the gardens was different in character and in the balance of species visiting. Each held the same core species, those that would be in the top-ten for pretty much any garden here in Britain, and each had a few differences. All of them, though, delivered the same opportunities to watch visiting birds and to learn more about their lives. Each offered up the same rewards and kept me connected with the natural world.

Touch wood, we'll be moving into a new home next year. Another rural garden and a property of great age and full of character. I am already excited by the prospect of what the garden bird community will be like – might I finally get tree sparrows visiting the garden feeding station? I hope that this will be my forever home, where several decades of watching will reward me with more wildlife encounters. I will continue to feed my garden birds and to take interest in them as individuals, despite my anxieties about the wider world. It is through doing this that I feel rooted in my locality, connected to the landscape and better able to understand what is happening on a broader scale. Like you, I am part of the natural world and I both have an influence on it through my activities and am influenced by it. The same is true for each of the birds that visits.

Acknowledgements

This book would not have been possible without Annabel Hill, whose love, support and kindness have shown me what a home should be. My parents, Basil and Mavis Toms, encouraged my interest in the natural world and were surprisingly tolerant of the activities of a budding naturalist, even when these involved the preparation of skulls or the housing of temporary invertebrate residents in my bedroom. They supported me through university and beyond, helping me to find my feet, and I miss their daily presence in my life.

I have been particularly fortunate to have spent my working life at the British Trust for Ornithology, a truly wonderful organisation that continues to deliver the ornithological knowledge and long-term data so essential for conservation. So many staff and former staff have helped me over the years and I am grateful to each and every one of them.

The ways in which I view the natural world have also been shaped by creative friends, whose artwork, poetry and written words continue to enrich my life and inspire. From the way in which Harriet Mead captures birds and animals in scrap metal, through the images evoked by the poetry of Matt Howard, and on to the wonderful writing of Mark Cocker, all lift my spirits, shape my thoughts and show me how richly rewarding the natural world can be for our own well-being.

Finally, I want to thank Myles Archibald, Hazel Eriksson, Tom Whiting, and all at HarperCollins, for their continued support of my work and for enabling me to share my experiences of the natural world with you.

Index

Mike Toms has overseen the Garden Ecology Team at the British Trust for Ornithology (BTO) for more than a decade, delivering research that addresses the question of why birds use gardens and the impact this use has on their ecology. Now Head of Communications at the BTO, Mike is responsible for how the organisation communicates the results of its scientific work to a broad range of audiences. Much of his work is geared towards public engagement in citizen science, delivering robust but accessible research through networks of keen volunteers.

Mike is a regular contributor to various magazines and the author of several books, including *Flight Lines*, *From Field & Fen*, *Gardening for Birdwatchers* and the Collins New Naturalist volumes *Owls* and *Garden Birds*. He is an advocate for the role that citizen science plays in supporting research and conservation, and is an active volunteer, passionate about the natural world and participating in a broad range of monitoring schemes.